COMPUTER EDUCATION
OF CHEMISTS

EDITED BY
PETER LYKOS
ILLINOIS INSTITUTE OF TECHNOLOGY

COMPUTER EDUCATION OF CHEMISTS

A WILEY-INTERSCIENCE PUBLICATION
JOHN WILEY & SONS
New York • Chichester • Brisbane • Toronto • Singapore

Based on a Symposium Presented
at the 184th National Meeting
of the American Chemical Society,
Kansas City, Missouri, September 1982

Copyright © 1984 by John Wiley & Sons, Inc.

All rights reserved. Published simultaneously in Canada.

Reproduction or translation of any part of this work
beyond that permitted by Section 107 or 108 of the
1976 United States Copyright Act without the permission
of the copyright owner is unlawful. Requests for
permission or further information should be addressed to
the Permissions Department, John Wiley & Sons, Inc.

Library of Congress Cataloging in Publication Data:
Main entry under title.

Computer education of chemists.

   Papers from a Symposium on Computer Education of
Chemists which was conducted under the auspices of the
ACS Divison of Computers in Chemistry.
   "A Wiley-Interscience publication."
   Includes bibliographies and index.
   Contents: Computer applications in chemistry / P.C.
Jurs — A college curriculum response to chemometrics /
R.P. DeToma — Information processing in chemistry / M.F.
Delaney and A.J. Antonitis — [etc.]

1. Chemistry—Data processing—Congresses. 2. Chemistry
—Data processing—Study and teaching—Congresses.
I. Lykos, Peter. II. American Chemical Society. Meeting
(184th : 1982 : Kansas City, Mo.) III. Symposium on
Computer Education of Chemists (1982 : Kansas City, Mo.)
IV. American Chemical Society. Division of Computers in
Chemistry.
QD39.3.E46C637   1984   540'.28'54   84-7528
ISBN 0-471-88872-9

Printed in the United States of America

10 9 8 7 6 5 4 3 2 1

CONTRIBUTORS

A. J. Antonitis, Boston University, Boston, Massachusetts
M. Brown, University of Virginia, Charlottesville, Virginia
M. F. Delaney, Boston University, Boston, Massachusetts
R. P. DeToma, Loyola College, Baltimore, Maryland
R. H. Heist, University of Rochester, Rochester, New York
P. C. Jurs, The Pennsylvania State University, University Park, Pennsylvania
M. G. Newton, University of Georgia, Athens, Georgia
T. Olsen, University of Rochester, Rochester, New York
G. F. Pollnow, University of Wisconsin-Oshkosh, Oshkosh, Wisconsin
K. L. Ratzlaff, University of Kansas, Lawrence, Kansas
H. Saltsburg, University of Rochester, Rochester, New York
J. W. Schilling, Trinity University, San Antonio, Texas
P. N. Swarztrauber, National Center for Atmospheric Research, Boulder Colorado
C. Trindle, University of Virginia, Charlottesville, Virginia
D. Zebolsky, Crieghton University, Omaha, Nebraska

PREFACE

In the fall of 1982, at the American Chemical Society National Meeting in Kansas City, there was a Symposium on Computer Education of Chemists which was conducted under the auspices of the ACS Division of Computers in Chemistry. At that symposium a number of persons, some not chemists, gave a snapshot in time of activities they were involved in regarding either introduction of computer-based chemical problem solving methods in chemistry curricula or aiding practicing chemists to learn some of those new methods that were relevant to their work.

In parallel with the events leading to that Symposium and continuing on was an activity of the American Chemical Society Committee on Professional Training regarding its approval program for bachelor degree programs in Chemistry whereby the Guidelines and Evaluation Procedures were being revised. That revision process included consideration of the range and depth of computer-based problem solving methods which should be part of undergraduate degree programs in chemistry. The revised guidelines were published in October 1983. In May 1984, ACS/CPT distributed a collection of course suggestions as appendices to the 1983 CPT guidelines. Of the first nine of these appendices, the three concerned with Analytical Chemistry, Chemical Information Retrieval, and Computers in Chemistry are all addressed by the chapters in this volume.

Accordingly it seemed timely to invite some of the participants in the ACS symposium to write, for publication, an account of their activity in that regard, including experiences and insights acquired since the symposium.

Their accounts of concrete experiences should be useful specifically to those who are concerned with adjusting their curricula in the light of the recently published ACS/CPT Guidelines and Evaluation Procedures, as well as to a much broader audience concerned with learning what some of the leaders, in a variety of situations, are doing in reaction to the impact of the computer on the practice of chemistry.

The entire range of computer science and engineering methodologies and technologies is being brought to bear on the practice of chemistry as well as methods of problem solving from other disciplines which have become feasible since the advent of the powerful microprocessor and its attendant interface/controller VLSI chips. Gathered here is a collection of accounts more appropriate for those with some familiarity with what is happening but who are concerned with introducing computers in the practice of chemistry rather than for those already at an advanced level.

Because Chemistry is an experimental science considerable attention is given to both the interfacing of computers to experimental apparatus and to the machine representation and transformation of data. On the other hand, the computer makes it possible to create models based on theory whereby data can be accommodated and new insights gained using numerical methods hitherto inaccessible to the chemist on any sort of reasonable scale. Chemistry is the physical science of bulkmatter viewed from a molecular perspective and so a major impact of the computer on chemistry is the ability to transform a three dimensional, simply connected, structure— a molecular diagram—to a string of characters for machine processing on the one hand and the ability to create a stereo pair of images of a molecular model on the other.

The chapters by Peter Jurs and Robert DeToma deal with a single course in a university setting on numerical and nonnumerical methods and with an entire college curriculum response to chemometrics, respectively. Michael DeLaney focuses on information processing in chemistry viewed in a rather broad context while Carl Trindle concerns himself with applications of algebraic symbol-manipulation.

FORTRAN has been the de facto standard computer language followed more recently by BASIC because of its wide use in personal computers. Computer scientists complain that both are obsolete and inappropriate, especially for beginners. The College Board's introduction of an advanced placement course in computer science starting in the fall of 1983 where the language of choice is PASCAL, has helped to promote further and wider use of PASCAL. With the better secondary school students entering college with programming skill in PASCAL, the work of Donald Zebolsky using PASCAL in a liberal arts university chemistry context demonstrates that languages other than FORTRAN and BASIC are workable in chemistry— and perhaps even preferable.

Preface ix

Paul Swarztrauber develops his experiences with helping chemists come to grips with nonstandard serial processor devices. Although his experience is with large scale scientific machines such as the CRAY 1, the low cost of microprocessors has already enticed chemists and other physical scientists to design and build highly concurrent multiprocessor machines that are task oriented and which require that algorithms be designed and coded taking into account the architecture of the device to be used. A major problem facing the chemistry establishment is that undergraduate and graduate students continue to be trained on serial computers, using FORTRAN and algorithms based on hand-calculator methods. Swarztrauber provides a window to the world of the nonserial processor in the service of chemistry.

There are four chapters dealing with the use of the computer in a laboratory environment. These range from Gilbert Pollnow's concern with the physical chemistry laboratory to Jess Schilling's dealing with juniors and seniors. Howard Saltsburg emphasizes hobbyist-level hardware and software in sophomore chemical engineering laboratory with the student in the control loop. Ken Ratzlaff addresses the issue from the perspective of one who recently joined a major university chemistry department for the purpose of exploiting computer science and the new small computer for all of chemistry.

This collection of papers spans the important ways that the computer is affecting chemistry other than situations where the hardware/software system is part of an applications package and the chemist can use the package without any deep knowledge of the supporting technology. Furthermore, through the authors' examples the interested reader has at one time a basis for calibration in terms of what can be done and some useful models which might be adapted and modified for one's own situation.

PETER LYKOS

Chicago, Illinois
August 1984

CONTENTS

1. Computer Applications in Chemistry: A University Course, 1

   P. C. Jurs

2. A College Curriculum Response to Chemometrics, 19

   R. P. DeToma

3. Information Processing in Chemistry, 47

   M. F. Delaney and A. J. Antonitis

4. Use of Algebraic Symbol-Manipulation Programs in Chemical Research and Education, 93

   C. Trindle, M. Brown, and M. G. Newton

5. Computer Applications in Chemistry, 108

   D. Zebolsky

6. Vector and Scalar Computers: A Comparison, 122

   P. N. Swarztrauber

7. Computer Science and the New Small Computer, 145

   K. L. Ratzlaff

8   The Microcomputer in the Undergraduate Laboratory, 154

   H. Saltsburg, R. H. Heist, and T. Olsen

9   Teaching Computers to Chemistry Students in a Liberal Arts University, 167

   J. W. Schilling

10 The MINC-11 as an Exemplary Laboratory Computer for Teaching and Research, 189

   G. F. Pollnow

Index, 221

COMPUTER EDUCATION OF CHEMISTS

PETER C. JURS   CHAPTER 1

Computer Applications in Chemistry: A University Course

Introduction

Computers have become an integral part of many aspects of modern life, including the conduct of science. Chemistry is no exception. The methods used in science have been profoundly changed by easy access to computational power that was unimaginable only a few years ago. As with other tools, computers affect chemistry in a variety of ways, some of which are obvious and some of which are hidden. For example, interaction with the Chemical Abstracts Service Online information retrieval service inherently requires some measure of computer knowledge, whereas the running of a modern Fourier Transform Infrared spectrometer does not, even though the instrument is fully automated. Thus, as computer technology has invaded chemistry, it has become incumbent upon chemical educators to produce a new generation of chemists who are literate in this area. It is essential that students have some knowledge of how computers impact on their own profession. Thus, we have developed and taught a course titled, "Computer Applications in Chemistry," at Penn State University over the past ten years. This paper will describe the course in detail.

2 Computer Education of Chemists

Goals

The two major subdivisions of computer applications in chemistry are (a) hardware interfacing and laboratory usages, where the computer becomes part of the instrument, and (b) software usage involving numerical calculations and abstract information-processing tasks. The primary goal of this course is to survey a variety of important computer software applications in chemistry. It is not feasible in one course to exhaustively enumerate all important applications, so a great deal of subjective selection has been done. We would like to introduce many topics in sufficient detail so that the students will be able to make further inquiries in a knowledgable way later in their careers and have the ability to read primary journal articles with confidence that touch on computer applications in chemistry.

Format

"Computer Applications in Chemistry" is a three semester-credit course. It is offered to juniors, seniors, and first-year graduate students as an optional course. The students come from majors such as chemistry, biochemistry, chemical engineering, and a smattering of others. It consists of lectures, discussions of problems and assignments, and a few demonstrations. The demonstrations are run on the Department of Chemistry PRIME 750 minicomputer. The students use the Penn State University Computation Center IBM 3033 for their assignments. All students are required to have a working knowledge of FORTRAN before starting the course.

Reading Material

The materials used during the course for reading assignments are listed in Tables I, II, and III. There is no single book that could serve as a textbook, but the five books listed in Table I are called optional books. Some of the students acquire them for their own libraries. A number of books are put on two-hour reserve in the library so that the students will all have ready access to them. Some assignments are made in these books, but they are largely used as reference materials. Photocopies or reprints of a selection of primary papers and reviews of pertinent literature are also put on two-hour reserve in the library. These papers are listed in Table III. As new material is published in the literature, the lists of books and papers changes and grows, and the lists are somewhat different for each offering of the course.

Computer Applications in Chemistry: A University Course

Course Topics

The course is divided into three parts: the introduction (15% of the time), numerical topics (50%), and nonnumerical topics (35%). The fraction of time spent on each part is approximate and varies somewhat from offering to offering.

Introduction

The topics covered in the opening portion of the course are meant to set the scene for what follows. The broad impact of computers on everyday life and science is discussed. The students read several articles from the popular press and descriptive reviews from Science (Simon 1977, Branscomb 1979, Davis 1977). Some elementary discussion of the internal organization of computers is provided. The key characteristics of modern digital computers that, when taken together, makes this particular piece of hardware uniquely capable are described. These are digital operation, stored program capability, self-controlling capability, automatic oepration, reliance on electronics, high speed, low error rate, and varied representation of information (integers, reals, codes for alphanumerics). Then an overview of scientific computer uses is given. These are listed as (a) numerical calculation or number crunching, (b) time-limited problems, (c) optimization, (d) information storage and retrieval, (e) experiment management and control, (f) intelligent problem solving, (g) visualization and graphical display, and (h) modeling and simulation. Under (a) the students are shown just how great the disparity between human computational power and computer is by calculating the number of operations per hour for a computer to be about the same as per lifetime for a person. Many examples suggest themselves, and some from outside chemistry are used as well as chemical ones. We discuss simulation as a source of new knowledge and artificial intelligence as well. A brief description of the information flow through a computer system is given, including the concepts of complication, assembly, editing, etc. The students must be shown how to interact with the specific computer systems they will use for the homework assignments.
Some concepts of program design and writing are covered next. The basics of structured programming are mentioned, and a number of specific tips on how to write efficient and more readable code are given. A number of the programming tips presented by Kernighan and Plauger (Kernighan 1974) are discussed. A few very specific code segments are given as illustrations of how to analyze problems -- for example, quick sorting and linearization of matrix subscripts. The basics of writing directive-driven programs are given.

4   Computer Education of Chemists

Error and statistics are introduced in the computational context. Accuracy and precision are defined. Error accumulation and how to estimate its effect are discussed. An excellent book by Bevington (Bevington 1969) is made available for the students. The fixed and floating point number systems are described, and their limitations are studied. The practical aspects of computer arithmetic are studied with the help of a technical note prepared by H. D. Knoble of the Penn State University Computation Center (Knoble 1979). The reasons why the laws of algebra break down in floating point computer calculations are given and discussed.

Numerical Topics

The numerical topics occupy the majority of the time of the course. The intent is to cover many subjects in sufficient detail so that the students will know the important concepts and know how to pursue the topic in more detail in the future if necessary. Several books are put on reserve in the library for the students to use during the numerical topics discussions (Beech 1975, Carnahan 1969, Johnson 1980, Pennington 1965).

The concept of an algorithm is introduced, and the criteria that are used to measure the goodness of an algorithm are discussed. The first section of numerical topics deals with the roots of equations. Direct methods are discussed as a starting point, and interative techniques follow. The requirements for an interative method to be useful are discussed in detail here because this is the first of many occurrences of iteration in the course. These requirements are (a) the means to make a satisfactory first guess, (b) the means to systematically improve on previous approximations, and (c) some criterion for stopping the iterations when sufficient accuracy is obtained. These abstract requirements are first applied to the bisection method of root finding, and then to the Newton-Raphson iterative method. The speeds of convergence and error properties of these methods are compared. The modified Newton method that can be applied to tangential roots is introduced as well. A homework assignment is given that applies the methods of root finding to the cubic equation that results from solving for the equilibrium hydrogen ion concentration and the pH of acetic acid solutions.

The second section under numerical methods deals with function generation. The point is made that digital evaluation of algebraic equations can be done based on many different functional forms, namely infinite series such as the Taylor series, continued fractions, rational function approximations, or other special algorithms. The tradeoffs in speed of execution and other properties are discussed. Evaluation of polynomials by Horner's rule is presented. To show in chemical terms what the power of a modern digital computer can do just in function evaluation, a homework assignment is given that develops

Computer Applications in Chemistry: A University Course 5

a potential energy surface for atom transfer reactions. A grid of potential energies is calculated, and the students must sketch equipotential lines on their output to form a contour map. A subroutine is handed out they can use to develop a sort of contour plot by shading using over-printing.

A large section of time is devoted to curve fitting. A book by Draper and Smith (Draper 1966) is available to the students. We start with the simplest, linear fit without weighting of points. Then linear fits with weights are introduced. Then higher order polynomial fits are discussed, using matrix notation for generality. Orthogonal polynomial fitting is covered. Transformations of functional form to help with fitting is discussed, and a homework assignment is given that involves fitting a set of kinetic data to get the preexponential factor and the activation energy. A second curve-fitting homework assignment involves the fitting of enzyme kinetic data to obtain the Michaelis-Menten constants. Three different approachs described in the literature are used and compared, and the students show that the most commonly used procedure is the worst one in terms of numerical stability. They are assigned a paper by Garfinkel that discusses curve fitting of biologically-oriented data in general and the various linear forms of the Michaelis-Menten equation in particular (Garfinkel 1980). Interative least squares fitting for complicated functional forms is covered, mathematical smoothing is introduced, and some examples from the literature of research involving curve fitting are shown. Finally, spline fitting is described.

The next topic in numerical analysis is integration. The trapezoidal rule and Simpson's rule are explained and compared for accuracy and speed. Gaussian quadrature is described. A homework assignment is given that involves the numerical integration of the gamma function using programs implementing Simpson's rule and Gauss-Laguerre quadrature. The two approaches are compared.

A section on matrix methods follows. The students taking this course have already had a mathematics course on matrix algebra, so we concentrate on numerical methods and their properties. We start with Cramer's rule for solution of systems of linear equations, and this leads into the necessity for evaluating determinants. Determinant evaluation by brute force and by making the determinant triangular using elementary row operations are discussed. The Gauss-Seidel iterative method for solution of linear equations is presented. The Gauss-Jordan elimination method for inversion of matrices or solution of linear equations is covered. A homework assignment that requires solving a four-by-four set of linear equations representing a mixture problem in infrared spectroscopy is given with a general matrix inversion subroutine that must be adapted to the problem.

6   Computer Education of Chemists

The next section is devoted to Monte Carlo methods. The basics are clearly presented by Fluendy (Fluendy 1970). We first discuss the concept of randomness and pseudorandom numbers. The commonly-used multiplicative method for generating pseudorandom numbers is presented in detail. A portable random number generator of Schrage (Schrage 1979) is discussed. Techniques for transforming uniformly-distributed random numbers to other distributions are presented, including inversion of the cumulative distribution function, the rejection technique, and other special methods. The term project involving a chemically-oriented version of the travelling salesman problem is presented at this time because the solution the students are asked to seek involves random numbers and simulation. Monte Carlo integration is briefly described. The application of random numbers to theoretical studies of polymer growth is presented.

The next numerical topic is numerical solution of differential equations. We discuss forward tracing methods in general, and then explore in detail Euler's method and the Runge-Kutta methods. The way in which the Runge-Kutta method can be applied to systems of first-order differential equations is shown using the notation of Carnahan, Luther, and Wilkes (Carnahan 1969). A homework assignment is given that involves solving a set of three first-order differential equations that models a set of simple chemical reactions. The students write a routine that implements the fourth-order Runge-Kutta method for a system of linked first-order differential equations to solve the problem. Finally, a predictor-corrector method is described using the particularly clear notation of Acton (Acton 1970).

During some offerings of this course there has been sufficient time to allow including a section on eigenanalysis. The perspective is taken that reduces the problem of finding eigenvalues and eigenvectors to the three-part problem of finding the characteristic polynomial, solving the characteristic polynomial for its roots, and solving sets of linear equations for the eigenvectors. The generation of characteristic polynomials is described by Pennington (Pennington 1965) as the Leverrier-Faddeev Method. Acton (Acton 1970) also has a nice discussion of eigenanalysis that develops the methodology in clear terms.

During some offerings of the course there has been sufficient time to include a section of the Fourier transform. The fundamentals of the Fourier series are developed using the approach of Stuart (Stuart 1961). Horlick (Horlick 1968) has also been used. A particularly short and simple FORTRAN fast Fourier transform subroutine presented by Mertz (Mertz 1971) is given to the students. The mathematics of the Fourier transform are tied to the Chemical world by discussing the Fourier transform infrared spectrometer in some detail using Horlick (Horlick 1968).

Computer Applications in Chemistry: A University Course 7

Nonnumerical Topics

The nonnumerical topics occupy the final third of the course. Once again, the intent is to cover a diversity of topics in enough detail so that the student will acquire the key concepts. They should learn enough to allow them to pursue the topics in greater detail later as the need arises.

The emphasis in this part of the course is on the abstract information-handling capabilities of digital computers. We discuss a number of ways computer software can aid chemistry where the emphasis is not on numerical computations.

The first topic discussed is chemical structure information handling (Lynch 1971). The degree to which chemists rely on drawn chemical structure diagrams is pointed out. These diagrams are really chemists' lowest common denominator regarding structural information and provide a means of visual communication. We them discuss the concepts of (a) linear notations, including a fairly detailed discussion of Wiswesser line notation (WLN), and (b) connection tables. WLN is presented with the help of two references: a paper by Gibson and Granito (Gibson 1972) and the reference book on WLN by Smith (Smith 1975).

WLN is presented by building up from simple structures to more complex ones by degrees, using many examples. The first dozen or so rules for generation of WLN from a compound's molecular structure are presented as they appear in Smith (Smith 1975). It is pointed out that WLN is a set of empirical rules that are undergoing change as they require updating.

The second major type of structural representation method discussed in detail is connection tables. This involves representing a structure by a matrix or set of matrices where the numerical entries represent bonds or atom types. A series of examples makes this simple system clear very quickly. It is shown how structures can be coded entirely as binary matrices, where the bonding pattern is contained in what has been termed an atom connectivity matrix. The direct link between chemical connection tables and the field of mathematics known as graph theory is presented. The primary purpose of our brief excursion into graph theory is to show that the connection table formalism has a mathematically rigorous underpinning, and that graph theory theorems can be exploited in chemistry, e.g., for ring perception in molecules, for registration, or for development of structural descriptors that can be correlated against physical properties.

The students are shown a working set of FORTRAN programs that support graphical entry of structures into computers. The programs are from the front end of our ADAPT (Automated Data Analysis using Pattern recognition Techniques) computer software system (Stuper 1979). The routine that supports the actual structural entry is named UDRAW (Brugger 1976) and is available from the Quantum Chemistry Program Exchange as program number 300 (Brugger 1976). The routine is written in

8   Computer Education of Chemists

standard FORTRAN and uses Tektronix graphics through calls to PLOT-10 software. It runs on our Chemistry Department PRIME 750 minicomputer. Moving it to other minicomputers with Tektronix graphics display terminals should not be too difficult. The program allows the user to sketch a structure on the display screen by using the keyboard and thumbwheel controllers for a cursor. All bookkeeping and construction of the connection table is done by the software. Thus, it demonstrates the power of digital computers to perform chemically meaningful, nonnumerical tasks.

A capability that is necessary in the development of large files of chemical structures is **registration**, that is, the ability to determine if a new compound being entered into the file is really new or matches a compound already present. If each compound is represented by a unique, unambiguous name, then registration is straight-forward and efficient. The Morgan algorithm is a method for generating a unique name, that is, connection table numbering, based on the graph theoretical properties of a compound's structure. It is explained clearly by O'Korn (O'Korn 1977) and an interesting extension including stereochemistry is presented by Wipke and Dyott (Wipke 1974).

The final topic taken up within chemical structure information handling is that of substructure searching. This too is related to the graph theory viewpoint because it is just an example of the graph isomorphism problem (Tarjan 1977).

The second major topic in the nonnumerical section of the course is molecular mechanics model building and graphical display. A useful reference for model building is Boyd and Lipkowitz (Boyd 1982) and for display Langridge et al. (Langridge 1981). The basic premise of molecular mechanics is that data determined experimentally for small molecules (bond lengths, bond angles, etc.) can be extrapolated to large molecules. A strain energy function is defined that is related to the positions of the atoms by terms that simultaneously attempt to optimize all bond lengths, bond angles, torsional angles, 1-3 and 1-4 interactions, etc. in the structure. This highly nonlinear strain energy equation is minimized by standard numerical techniques, and the resulting models are displayed graphically. The ADAPT system contains a program that performs these computations, and it is executed for the students to show the procedures as they are actually done. A space-filling display of the models is generated to show how computer graphics can enhance visualization of chemical structures. Figure 1 shows the molecule norbornone displayed several different ways. These drawings were generated by a version of the program reported by Smith and Gund (Smith 1978).

The topic of **optimization** is discussed from the point of view of simplex optimization (Deming 1973, Deming 1974, Shavers 1979). The ideas of evolutionary operation and the simplex are described. These are methods for determining the optimal values for a set of variables which are influencing an observable

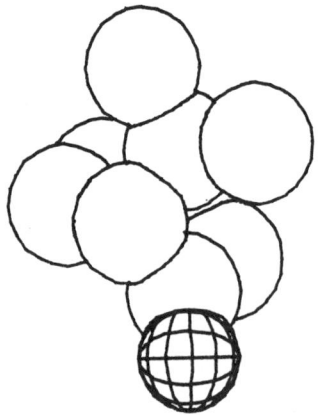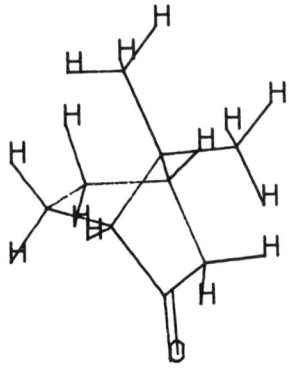

Figure 1. Four Views of Norbornone.

dependent variable. Applications of simplex optimization to gas chromatography (Morgan 1975) and flame spectroscopy (Routh 1977) are discussed. The tie between these experimental implementations of simplex optimization and the more general field of linear programming (or operations research) is suggested.

The topic of pattern recognition is introduced along with some chemical examples (Jurs 1975, Kowalski 1975). The students are shown how the notation of pattern recognition can be used to represent many kinds of chemical data, and how the methods can be used to investigate large sets of data for similarities. The subdivisions of pattern recognition -- preprocessing, mapping and display, development of discriminants, and clustering are discussed. It is pointed out that quite a variety of applications have appeared in the chemical literature, including spectral analysis, materials analysis, and structure-property and structure-activity relations. Several applications papers are discussed in class: classification of crude oils (Clark 1979) and classification of archaeological materials (McGill 1977).

A nonnumerical topic that is only loosely related to chemistry, but is very popular with the students, is artificial intelligence (AI). From the very beginning of the course and the introductory readings, intruiging questions are posed regarding the possibility of machine intelligence and whether computer software can generate new knowledge. Two recent, outstanding books in this area are recommended to the students, but not assigned in this course -- the Pulitzer Prize winner Godel, Escher, Bach: the Eternal Golden Braid (Hofstadter 1979) and The Sciences of the Artificial (Simon 1969). An article from The New York Times covering AI in a nontechnical way is assigned (Stockton 1980). Some of the fundamental ideas of AI research are presented such as Turing's "imitation game," the idea of heuristics, and the basis of expert systems. An article reviewing AI expert systems is assigned (Duda 1981). The basic properties of expert systems are explained, and the ideas are illustrated using the chemical example DENDRAL AI project in the area of computer-assisted structure elucidation. The heart of the project is the program CONGEN (for constrained structure generation) that can provide all chemical structures consistent with user-supplied constraints (Carhart 1977). Many additional AI systems ouside of chemistry could also be used as examples.

A nonnumerical computer topic that impacts on all chemists is chemical information retrieval. This subject is briefly covered to introduce the students to a few of the basic concepts. The subunits of an information retrieval system and the types of files are covered. Table lookup techniques are discussed, and hash coding is described (Wipke 1978). We have a brief discussion of some of the chemical information retrieval services that deal with spectroscopic data (Heller 1980).

Discussion

"Computer Applications in Chemistry" has been offered over a period of more than ten years. The contents, readings, and assignments have evolved as topics have been added and deleted from the course. New reading assignments and reference articles are added to the reading lists as they appear. The students have used a batch-oriented Computation Center computer for most of the offerings, but recently there has been a change to a fully time-sharing system, so the course has moved to it. The enrollment has been in the range of 25 to 35 students per offering, with most of the students coming from chemistry, computer science, and biochemistry. A few first-year graduate students also take the course. The students have been asked to rate the course as to timeliness, utility of material for their future careers, and so on. The responses are largely positive, and the students evidently perceive the material to be in their interest. The students do recommend the course to their peers, and the enrollment is holding steady, so that course seems to fill a need in our chemistry curriculum.

References

F. S. Acton, "Numerical Methods that Work," Harper and Row, New York, 1970.

G. Beech, "FORTRAN IV in Chemistry," Wiley, New York, 1975.

P. R. Bevington, "Data Reduction and Error Analysis for the Physical Sciences," McGraw-Hill, 1969.

D. B. Boyd and K. B. Lipkowitz, "Molecular Mechanics. The Method and Its Underlying Philosophy," Jour. Chem. Ed., 59, 269-274 (1982).

L. M. Branscomb, "Information: The Ultimate Frontier," Science, 203, 143-147 (1979).

R. Bracewell, "The Fourier Transform and its Applications," McGraw-Hill, New York, 1965.

W. E. Brugger and P. C. Jurs, "Molecular Structure Input Program Using a Storage Cathode Ray Tube Terminal," Anal. Chem., 47, 781-784 (1975).

W. E. Brugger and P. C. Jurs, "UDRAW: Molecular Structure Input Program Using a CRT Terminal," QCPE Program No. 300, Quantum Chemistry Program Exchange, Department of Chemistry, Indiana University, Bloomington, Indiana 47907.

R. A. Carhart, T. H. Varkony, D. H. Smith, "Computer Assistance for the Structural Chemist," A. C. S. Symp. Ser., 54, 126-145 (1977).

B. Carnahan, H. A. Luther, J. O. Wilkes, "Applied Numerical Methods," John Wiley & Sons, Inc., New York, 1969.

H. A. Clark and P. C. Jurs, "Classification of Crude Oil Gas Chromatograms by Pattern Recognition Techniques," Anal. Chem., 51, 616-623 (1979).

R. M. Davis, "Evolution of Computers and Computing," Science 195, 1096-1102 (1977).
S. N. Deming and S. L. Morgan, "Simplex Optimization of Variables in Analytical Chemistry," Analytical Chemistry, 45, 278A-283A (1973).
N. R. Draper and H. Smith, "Applied Regression Analysis," Wiley, New York, 1966.
R. O. Duda and J. G. Gaschnig, "Knowledge-Based Expert Systems Come of Age," Byte, Sept. 1981, pp. 238ff.
C. G. Enke, "Computers in Scientific Instrumentation," Science, 215, 785-791 (1982).
M. Fluendy, "Monte Carlo Methods," in "Markov Chains and Monte Carlo Calculations in Polymer Science," G. G. Lowry (Ed.), Marcel Dekker, Inc., New York, 1970, pp. 45-90.
David Garfinkel, "Computer Modeling, Complex Biological Systems, and their Simplifications," Amer. Jour. Physiol., 239, R1-R6 (1980).
G. W. Gibson and C. E. Granito, "Wiswesser Chemical Line-Notation," American Laboratory, 4, 27-37 (1972).
S. R. Heller and G. W. A. Milne, "On-line Spectroscopic Data Bases," Amer. Lab., March 1980, pp. 33ff.
D. R. Hofstadter, "Godel, Escher, Bach: An Eternal Golden Braid," Basic Books, 1979.
G. Horlick, "Introduction to Fourier Transform Spectroscopy," Appl. Spectr., 22, 617-626 (1968).
K. J. Johnson, "Numerical Methods in Chemistry," Marcel Dekker, Inc., New York, 1980.
P. C. Jurs and T. L. Isenhour, "Chemical Applications of Pattern Recognition," Wiley-Interscience, New York, 1975.
B. W. Kernighan and P. J. Plaugher, "The Elements of Programming Style," McGraw-Hill, New York, 1974.
H. D. Knoble, "A Practical Look at Computer Arithmetic. A Three-Part Tutorial," Computation Center Technical Note, Computation Center, The Pennsylvania State University, University Park, PA 16802, 1979.
B. R. Kowalski, "Measurement Analysis by Pattern Recognition," Anal. Chem., 47, 1152A (1975).
R. Langridge, T. E. Ferrin, I. D. Kuntz, M. L. Connolly, "Real-Time Color Graphics in Studies of Molecular Interactions," Science, 211, 661-666 (1981).
M. F. Lynch, J. M. Harrison, W. G. Town, "Computer Handling of Chemical Structure Information," Macdonald, London, 1971.
J. R. McGill and B. R. Kowalski, "Recognizing Patterns in Trace Elements," Appl. Spectr., 31, 87-95 (1977).
L. Mertz, "Fourier Spectroscopy: Past, Present, and Future," Appl. Optics, 10, 386 (1971).
S. L. Morgan and S. N. Deming, "Simplex Optimization of Analytical Chemical Methods," Analytical Chemistry, 46, 1170-1181 (1974).

L. J. O'Korn, "Algorithms in the Computer Handling of Chemical Information," in R. E. Christoffersen (Ed.), "Algorithms for Chemical Computations," A. C. S. Symp. Ser., 46, 122-148 (1977).

R. H. Pennington, "Introductory Computer Methods and Numerical Analysis," MacMillan, 1965.

M. W. Routh, P. A. Swartz, M. B. Denton, "Performance of the Super Modified Simplex," Analytical Chemistry, 49, 1422-1428 (1977).

Linus Schrage, "A More Portable FORTRAN Random Number Generator," A. C. M. Transactions on Mathematical Software, 5, 132-138 (1979).

C. L. Shavers, M. L. Parson, S. N. Deming, "Simplex Optimization of Chemical Systems," Jour. Chem. Ed., 56, 307-309 (1979).

H. A. Simon, "The Sciences of the Artificial," MIT Press, 1969.

Herbert A. Simon, "What Computers Mean for Man and Society," Science, 195, 1186-1191 (1977).

E. G. Smith, "The Wiswesser Line-Formula Chemical Notation," McGraw-Hill, New York, 1975.

G. M. Smith and P. Gund, "Computer-Generated Space-Filling Molecular Models," Jour. Chem. Inf. Comput. Sci., 18, 207-210 (1978).

William Stockton, "Creating Computers That Think," New York Times, Dec. 7, 1980.

R. D. Stuart, "An Introduction to Fourier Analysis," Methuen and Co., 1961.

A. J. Stuper, W. E. Brugger, P. C. Jurs, "Computer Assisted Studies of Chemical Structure and Biological Function," Wiley-Interscience, New York, 1979.

R. E. Tarjan, "Graph Algorithms in Chemical Computations," in R. E. Christoffersen (Ed.), "Algorithms for Chemical Computations," A. C. S. Symp. Ser., 46, 1-20 (1977).

W. T. Wipke and T. M. Dyott, "Stereochemically Unique Naming Algorithm," Journal of the American Chemical Society, 96, 4834-4842 (1974).

W. T. Wipke, S. Krishnan, G. I. Ouchi, "Hash Functions for Rapid Storage and Retrieval of Chemical Structures," Jour. Chem. Inf. Comput. Sci., 18, 32-37 (1978).

14 Computer Education of Chemists

Table I
Textbook List

T. L. Isenhour and P. C. Jurs
Introduction to Computer Programming for Chemists
   2nd Edition
Allyn and Bacon, 1979.

G. Beech
FORTRAN IV in Chemistry
John Wiley and Sons, 1975.

K. J. Johnson
Numerical Methods in Chemistry
Marcel Dekker, Inc., 1980.

A. C. Day
FORTRAN Techniques with Special References to
   Nonnumerical Techniques
Cambridge University Press, 1972.

B. W. Kernighan and P. J. Plauger
The Elements of Programming Style
McGraw-Hill, 1974.

Table II
Books on Two-Hour Reserve

Introduction to Computer Programming for
  Chemists: FORTRAN, 2nd Ed.
Allyn and Bacon, 1979.

B. R. Kowalski (Ed.)
Chemometrics: Theory and Application
A. C. S. Symposium Series, Vol. 52, 1977.

R. N. Bracewell
The Fourier Transform and Its Applications
McGraw-Hill, 1965.

K. J. Johnson
Numerical Methods in Chemistry
Marcel Dekker, 1980.

W. T. Wipke and W. J. Howe (Eds.)
Computer-Assisted Organic Synthesis
A. C. S. Symposium Series, Vol. 61, 1977.

D. H. Smith (Ed.)
Computer-Assisted Structure Elucidation
A. C. S. Symposium Series, Vol. 54, 1977.

R. E. Christoffersen (Ed.)
Algorithms for Chemical Computations
A. C. S. Symposium Series, Vol. 46, 1977.

G. Beech
Fortran IV in Chemistry
Wiley, 1975.

K. N. Rao (Ed.)
Molecular Spectroscopy: Modern Research, Vol. 2
Academic Press, 1976.

Committee on Chemical Information N. R. C.
Chemical Structure Information Handling
National Academy of Sciences, 1969.

Table II Continued

M. F. Lynch, J. M. Harrison, W. G. Town
Computer Handling of Chemical Structure Information
Macdonald, 1971.

E. G. Smith
The Wiswesser Line-Formula Chemical Notation
McGraw-Hill, 1975.

A. T. Balaban (Ed.)
Chemical Applications of Graph Theory
Academic Press, 1976.

R. H. Pennington
Introductory Computer Methods and Numerical Analysis
MacMillan, 1965.

B. Carnahan, H. A. Luther, J. O. Wilkes
Applied Numerical Methods
Wiley, 1969.

B. W. Kernighan and P. J. Plaugher
The Elements of Programming Style
McGraw-Hill, 1974.

N. R. Draper and H. Smith
Applied Regression Analysis
Wiley, 1966.

Peter Lykos and Isaiah Shavitt (Eds.)
Supercomputers in Chemistry
A. C. S. Symposium Series, Vol. 173, 1981.

P. R. Bevington
Data Reduction and Error Analysis for
   the Physical Sciences
McGraw-Hill, 1969.

A. C. Norris
Computational Chemistry
Wiley, 1981.

U. Burkert and N. L. Allinger
Molecular Mechanics
Amer. Chem. Soc., 1982.

K. Varmuza
Pattern Recognition in Chemistry
Springer-Verlag, Berlin, 1980.

Computer Applications in Chemistry: A University Course

Table III
Articles on Two-Hour Researve

S. N. Deming and S. L. Morgan
Simplex Optimization of Variables in Analytical Chemistry
Analytical Chemistry 45:278A-283A (1973).

Herbert A. Simon
What Computers Mean for Man and Society
Science 195:1186-1191 (1977).

S. L. Morgan and S. N. Deming
Simplex Optimization of Analytical Chemical Methods
Analytical Chemistry 46:1170-1181 (1974).

L. M. Branscomb
Information: The Ultimate Frontier
Science 203:143-147 (1979).

M. W. Routh, P. A. Swartz, M. B. Denton
Performance of the Super Modified Simplex
Analytical Chemistry 49:1422-1428 (1977).

S. R. Heller and G. W. A. Milne
On-line Spectroscopic Data Bases
American Laboratory, pp. 33-48, March 1980.

R. M. Davis
Evolution of Computers and Computing
Science 195:1096-1102 (1977).

William Stockton
Creating Computers That Think
New York Times, December 7, 1980.

The Computer Revolution
Chemical and Engineering News, 8/21/78, p. 5.

G. W. Gibson and C. E. Granito
Wiswesser Chemical Line-Notation
American Laboratory 4:27-37 (1972).

W. T. Wipke and T. M. Dyott
Stereochemically Unique Naming Algorithm
Journal of the American Chemical Society 96:4834-4842 (1974).

Table III Continued

R. Langridge, T. E. Ferrin, I. D. Kuntz, M. L. Connolly
Real-Time Color Graphics in Studies of Molecular Interactions
Science 211:661-666 (1981).

C. G. Enke
Computers in Scientific Instrumentation
Science 215:785-791 (1982).

David Garfinkel
Computer Modeling, Complex Biological Systems, and
   their Simplifications
Amer. Jour. Physiol. 239:R1-R6 (1980).

Donald B. Boyd and Kenny B. Lipkowitz
Molecular Mechanics. The Method and Its Underlying Philosophy
Jour. Chem. Ed. 59:269-274 (1982).

W. T. Wipke, S. Krishnan, G. I. Ouchi
Hash Functions for Rapid Storage and Retrieval of
   Chemical Structures
Jour. Chem. Inf. Comput. Sci. 18:32-37 (1978).

R. O. Duda and J. G. Gaschnig
Knowledge-Based Expert Systems Come of Age
Byte, September 1981, pp. 238ff.

John J. Vollmer
Wiswesser Line Notation: An Introduction
Jour. Chem. Education, 60, 192 (1983).

ROBERT P. DeTOMA CHAPTER 2

A College Curriculum Response
to Chemometrics

INTRODUCTION

It is currently recognized that the time has come for the formal inclusion of chemical applications of mathematical techniques (chemometrics) and computerized methods in the undergraduate chemistry curriculum.[1] This direction toward chemometric and computer methods has certainly been stimulated by the increasing availability and capabilities of computers. Modern computer-based instrumentation allows for the acquisition, processing, and storage of large quantities of high fidelity data. These capabilities and the natural digital representation of information they provide lead to the development of new experimental design concepts not thought possible several years ago. Chemometrics is the tool for extracting optimal chemical information from chemical data. Its application can be as diverse as confirming the presence of a dilute component in a spectrally overlapping solvent matrix by ir spectroscopy (which might involve the mathematical techniques of ensemble averaging, and $-\log x$, $-$, 10^{-x} operations on spectral data arrays) to the mathematical design concepts embodied in multiplex spectroscopy or the multidimensional analysis of complex data structures. This rapidly growing field includes such topical areas as multivariate statistics, simulation, experimental design optimization, pattern recognition, to name a few - all of which make it possible to view, analyze and interpret chemical information and chemical problems.

The world of modern chemical practice relies heavily on the computer/chemometric approach to problem solving. It would appear

that all chemistry students entering this world should be brought
to a threshold level of awareness and practice in these methods
if they are going to perform effectively in it. The immediate
problem faced by a chemistry department wishing to implement this
training is how to fit it into an already over-crowded curriculum.
This problem is especially pronounced in an undergraduate liberal
arts environment where the number of courses that can be offered
in a major has a definite limit.

In this paper a strategy is presented whereby computer/chemometric techniques can be formally incorporated into any undergraduate chemistry curriculum. The basic approach is to enhance (not degrade by overemphasis) existing chemistry courses by blending into them the new material on computer/chemometric methods where appropriate. It must be recognized that full implementation of such a plan will not take place overnight. Since this approach will involve a variety of chemistry courses, careful organization and coordination will be required. In addition, time will have to be allowed for faculty development in the various areas, since it is currently unlikely that a sufficient sample of computer/chemometrics expertise will exist in a small chemistry department.

COMPUTER/CHEMOMETRICS PROGRAM STRATEGY

The problem of curriculum expansion associated with the teaching of computer/chemometric methods will be discussed in the context of a four-year liberal arts institution, specifically taking Loyola College as an example. The overall curriculum at Loyola consists of three parts: core courses, major courses, and elective courses. The core is required of all students regardless of major and is intended to introduce students to the areas listed in Figure 1. This liberal arts framework restricts the number of courses that can be offered in a major.

Before the inclusion of any computer/chemometric enhancements, the ACS certified course program in chemistry at Loyola filled the curriculum quota, leaving no room for new four-credit courses in the major (Loyola is on a 4-1-4 system). One redundancy did, however, exist in the program in connection with Physical Chemistry Laboratory which met twice a week, four hours per meeting for two semesters. The original purpose of this "double" laboratory exposure was to provide instrumentation experience when the Instrumental Methods course did not exist several years ago. Thus, a two-credit laboratory slot can be opened up by delating the extra physical chemistry laboratory without any serious loss. But this is hardly adequate to provide the level of computer/chemometrics instruction desired. The situation at Loyola is that there is very little room available for real curriculum expansion by way of introducing new courses.

In attempting to develop a strategy that would provide formal instruction in computer/chemometric methods, it was recognized that a focal course in chemometrics with emphasis on numerical methods and applied statistics was desirable in order

A College Curriculum Response to Chemometrics

```
CORE  +  MAJOR  +  ELECTIVES
  ‖
history (2)

ethics (1)

language (2)

literature (2)

math-sci (3)

philosophy (2)

social sci (2)

theology (2)

composition (1)
```

Figure 1.
Overall curriculum at Loyola College. Numbers in parentheses indicate the number of four-credit courses required in each area.

to provide a degree of structure to the overall program. It was also recognized that any valid course in chemometrics should require a certain level of computer programming expertise, otherwise the chemometrics course turns into a programming course or a course in using other people's "black boxes" without understanding. Thus, already it appears that two new courses, chemometrics and a prerequisite programming course, are desirable and the important topics of laboratory and non-numerical computing have yet to be considered.

In addition to the space problem noted above, the problem of where to locate the new chemometrics course in the curriculum logically follows. On the one hand, it would appear that this course should be offered late in the program (say senior year) so that students will have enough chemical background to appreciate the breadth of coverage chemometrics offers. But, on the other hand, if the training comes this late, will students get enough practice? It would indeed be of value if students could use computer and chemometric techniques to solve problems they encounter in their regular chemistry courses and laboratories when it is meaningful to do so. Certainly, ample opportunity exists for using computer/chemometric techniques throughout the four-year program. If these areas are treated as basic chemical tools, as are wet analytical methods or stoichiometry, then it is possible to take advantage of this opportunity. Wet analytical methods and stoichiometry are encountered many times in an undergraduate chemistry program and each time a student uses them again he/she develops more skill in their use, all in a very natural way. This same approach can be used in teaching chemometrics and computer methods, ie. instead of trying to cram everything into a single course normally taken (if at all) in the senior year, introduce material early and let it develop naturally and gradually. In each stage of development, a level of application understandable to the student can be presented allowing the training to grow and build on itself. Where laboratory is involved, experiments can be selected to achieve a balance between traditional and computer/chemometric-assisted approaches which in fact simulates methodologies used in modern chemical practice.

At Loyola College this "aufbau" approach to teaching chemometrics and computer methods has been adopted. The aufbau approach provides a mechanism where new material can be introduced into a curriculum that will not allow for the significant addition of new courses. In addition, it provides a mechanism whereby computer/chemometric methods can be practiced thus enabling a real level of experience to develop. The aufbau plan at Loyola involves one specifically computer/chemometrics oriented course and a number of standard chemistry courses which provide background for, use, and build on this specific training as appropriate.

The courses directly involved in this program are listed in Table 1. The training begins early in the sophomore year with Quantitative Analysis and ends late in the senior year with Instrumental Methods. Quantitative Analysis serves as a foundation course in computer programming and also provides an introduction to the use of computers in chemistry. The only new course in the curriculum is Computational Methods in Chemistry Lab (CMC Lab) which replaces the physical chemistry laboratory noted earlier. CMC Lab involves 3-4 hours of lecture plus 4-8 hours of project work per week. This is the same amount of time that students would spend in a physical chemistry laboratory experience. CMC Lab represents the focal course in the program and is located midstream (junior year) to take advantage of the mathematical and computer skills developed in the first two years as well as to allow practice experience of computer/chemometric methods in the remaining junior year and senior year. The remaining courses in Table 1 use and quietly develop computer/chemometric methods.

In the original plan of this program a special 2-credit course in BASIC programming was to be taken concurrently (and designed to interact) with Quantitative Analysis. This course was to serve as the computer programming prerequisite for CMC Lab noted earlier. Due to staffing problems in the Computer Science Department this special 2-credit course could not be offered when the program was first implemented (1981-82 academic year). Rather than introduce a new programming course in the Chemistry Department, it was decided to provide this experience

A College Curriculum Response to Chemometrics

Table 1

A Summary of Courses Directly Involved in
Computer/Chemometrics Training

COURSE (CREDITS)	SEMESTER	YEAR
Quantitative Analysis (4)	1st	Sophomore
Computational Methods in Chemistry Lab (2)*	1st	Junior
Physical Chemistry I (4)	1st	
Physical Chemistry II (4)	2nd	
Physical Chemistry Lab (2)	2nd	
Qual Organic Lab (2)	1st	Senior
Topics in Advanced Chemistry (4)	2nd	
Instrumental Methods (4)	2nd	

*New Course

---

directly in the Quantitative Analysis course. The details of doing this without normal course material loss will be discussed in the next section. Also, at the time of the original program plan, the second semester General Physics Laboratory taken in the junior year was essentially a laboratory course in digital electronics and fundamental interface design. This course was to provide a foundation for chemical instrumentation interfacing practice in the senior year Instrumental Methods course. It was indeed a disappointment when the Physics Department decided to change the direction of this laboratory and no longer emphasize digital electronics. Physical Chemistry Laboratory and Instrumental Methods will now have to share the responsibility of providing a minimum operational background in common digital circuit components and subsystems. These are examples of the kinds of logistical problems encountered when attempting to develop a program of this type.

Before presenting details of the degree of computer/chemometrics involvement associated with each year of the Loyola program, it is important to consider the underlying theme or emphasis of this program. When this project was conceived several years ago (1980), one of our major goals was to teach students how to operate (with understanding and competence) in a modern laboratory environment equipped with computer-based instrumentation. This includes off-line, on-line, and in-line applications. We wanted to get students to recognize that spectra, chromatograms, titration curves, etc. are data structures that represent mathematical functions that can be manipulated and operated on mathematically to derive useful chemical information; and to recognize that if these data were only represented in analog form that the chemometric approach and its benefits would not be practical or feasible. Students should understand what that in-line "black box" is doing to their data. They should be able to assemble an algorithm they might need that is not available in an instrument's firmware. They should be able to connect an instrument equipped with a standard digital I/O port (RS232C, BCD, IEEE-488) to a general purpose laboratory computer and be able to write the communications software. Students should know how to logically approach an interfacing problem that involves A/D or D/A conversion, event timing, event control, interrupt programming, etc. The picture that emerges here is clear. Our feeling was/is that much of this kind of experience is generally useful to all chemists who will use and derive benefit from modern computer-based instrumentation and not just to potential analytical or physical chemists. Thus, at Loyola, we felt a strong need to start our computer/chemometrics teaching program by emphasizing data treatment and acquisition. There is sufficient flexibility in the "aufbau" structure of the program to allow for upgrading and refinement once the initial implementation gets underway. For example, non-numerical methods like pattern recognition, library searching, and computer-assisted organic synthesis can be blended into other courses like Organic Chemistry, Inorganic Chemistry, and Biochemistry. This "blending" process can be accomplished in such a way that preserves the normal course material by first giving an exposure lecture on the algorithm, followed by assigned readings on the algorithm details, and finally by using and/or modifying the algorithm in assigned projects. Perhaps one such topic might be assigned to each course. If properly developed and organized, it appears that this approach to teaching computer/chemometric methods can only enrich the curriculum.

COMPUTER/CHEMOMETRICS PROGRAM DETAILS

It should be noted that the computer/chemometrics teaching program at Loyola was first implemented in the 1981/82 academic year and is still under development. Time is needed to attend to the many necessary hardware/software details as well as for faculty development. Because the program involves the sophomore, junior and senior years of a given class it will take three years to complete one cycle and therefore also to arrive at a minimal evaluation. In this section some of the details of the program

A College Curriculum Response to Chemometrics

will be considered. In addition some results from the first years implementation will be presented.

Sophomore Year

Table 2 summarizes the courses and goals relevant to computer/chemometrics training in the sophomore year. The overall objectives in this year are to expose students to a computational system and to the use of such a system to solve some chemical problems. The only course involved is Quantitative Analysis. In one aspect of this course, students will learn simple programming technique in BASIC as well as become familiar with the college computer operating system utilities and periferals (file manipulation, text editor, line-printer, graphics terminal). This experience will serve to remove the aura of mystery normally associated with the computer and allow students to start becoming comfortable with its use.

A valid question at this point is how can the above be accomplished without sacrificing normal subject matter in the Quantitative Analysis course. Based on two years experience with an approach that answers this question, not only can it be accomplished, but it can be accomplished in such a way that it enhances the normal subject matter of the course. The approach is based on the slow and homogeneous diffusion of the computer-related material throughout the course. The teaching vehicle is by example, where the examples are taken directly from the Quantitative Analysis course, and through carefully designed assignments which introduce the computer material at a graded level of sophistication. During the early part of the course several introductory lectures on computers and programming are given in the first hour of a normal laboratory session. These lectures are sufficiently staggered so that students have enough time to work their programming assignments which are developing their programming skills. At this stage computer applications are simple function evaluations which are used to develop the programming techniques of conditional branching, looping, I/O, data types, subscripted variables, etc. Then when a nice chemical application presents itself in lecture it can be followed (one or two weeks later to allow the chemistry to sink-in first) with a computer assignment based on the chemical application. This assignment, in addition to reinforcing previously learned programming, should be designed to include new programming concepts so that the level of programming skill can be extended with each new assignment. It is important that assignments be regular in order to provide a sufficient degree of practice experience that is necessary in learning computer programming. In addition, this programming aspect of the course should be regarded as serious and student performance in this area should be evaluated in normal course examinations.

Each new programming assignment might require a 15-30 minute lecture to introduce; this, together with the previously noted introductory programming lectures suggests that the actual time

Table 2

Computer/Chemometrics Related Courses and Goals

YEAR	COURSES/—GOALS
Sophomore	Quantitative Analysis —Introduction to a computer system —Introduction to computer programming —Introduction to computer/chemometrics applications, theory —Introduction to computer/chemometrics applications, laboratory —Exposure to computer-assisted experimentation
Junior	Computational Methods in Chemistry Lab —Computer programming skills refinement —Chemometrics (numerical methods and applied statistics emphasis) Physical Chemistry —Computer/chemometrics applications; problem solving, simulation Physical Chemistry Lab —Use of computer-assisted experimentation —Computer/chemometrics applications, laboratory —Introduction to digital electronics
Senior	Qualitative Organic Analysis Lab —Library/Literature searching Topics in Advanced Chemistry —The eigenvalue-eigenvector problem —Differential equations, numerical methods Instrumental Methods —Interface design/practice with chemical instrumentation —Use of computer-assisted experimentation —Computer/chemometrics applications, laboratory —Simulation experiments

spent teaching computer programming represents a small perturbation in the overall course scheme which can easily be absorbed in the laboratory segment of the course. The bulk of the work is done by the students in the assignments. For this reason the assignments must be simple enough so that they are not a burden to the students, yet also carry a sufficient degree of rigor so that the students learn some programming.

The source of assignments in the Quantitative Analysis course is extensive; for example: alpha-distribution generation and graphics, small sample statistics, exact equilibrium calculations, piecewise and global titration curve generation, distribution coefficient profile modeling relevant to liquid-liquid extraction equilibria, chromatography elution profile simulation, to name a few. When applications require the introduction of numerical and statistical methods (like root-finding or least squares estimation) it has been found expedient to simply present the algorithm with only a brief but understandable justification of its origin. Rigorous theoretical development of algorithms can be saved for CMC Lab. The object here is to teach computer programming by way of using algorithms and functions of relevance to chemistry. Any algorithm that is felt to be too sophisticated for students to handle at this level (eg., a graphics display routine, or a non-linear simultaneous equation solver) can be formulated as a subroutine by the instructor and given to the students for use in their programs. This approach not only renders a complex problem manageable for the students, but it also illustrates to them a very important concept in structured programming technique as well as the general utility that a subprogram formulation provides. One other consideration is worth noting here. Not only do these computer-related assignments teach computer programming, but they also illustrate chemistry.

Emphasis in the laboratory segment of Quantitative Analysis at Loyola is on wet techniques and non-instrumentally enhanced separations. However, there are several excellent opportunities in this laboratory where chemometric methods can be used in data reduction and where computer-controlled experimentation can be illustrated. For example, two experiments that are now done, one involving a potentiometric titration and the other, a photometric titration, are carried out as normal using manual instrument control, manual data acquisition and graphical analysis of the data. Data files of these raw titration curve data are then created manually through a keyboard (entering data into the computer by hand serves as a nice introduction to the computer system's text editor). These data are then analyzed by numerical and statistical techniques which have been coded to operate interactively and with graphics enhancement so that students can see what they are doing to their data at any stage of the processing. The photometric experiment involves the location of an intersection and its confidence interval. An algorithm which employs two-parameter linear least squares fitting with iterative point dropping and sample variance monitoring works well for this analysis and is readily understood by the students. In the analysis of the potentiometric data, a two-pass algorithm was developed based on local polynomial regression to locate an inflection by derivative techniques. In the first pass a local polynomial segment is moved through and fitted to the titration data producing an interpolated first derivative of these data. The proper choice of polynomial degree and local segment extent depends on data density and quality, and thus requires students to exercise judgement based on the quality of the fit to the data. The second pass uses the first derivative as the input file and by the same local fitting method produces the second derivative of the titration curve and thus locates the end point at its dominant zero. Figure 2 illustrates the method with data taken from a student titration.

When computer-controlled versions of these same two experiments are implemented, they will be demonstrated to the students. Their design will incorporate closed loop control and a "live" display that is updated with each newly acquired datum. These experiments will directly produce the final titration end point and its confidence interval (as opposed to an off-line interactive approach) and will output the raw high density and high quality titration curve data to a file which can be accessed by students for careful visual examination with a high resolution graphics device. The results will then be discussed in light of a comparison of the computer/chemometric enhanced and manual approaches.

This project serves several useful purposes. Interactive chemometric techniques are used by students to analyze their own data, thus a special degree of attention is attached to the learning process. Students quickly realize the enhancements provided by chemometric methods as well as the possibility of deterioration if these methods are not properly used or understood. In addition, students are exposed to computer-aided experimentation and to the ease and quality of large sample data handling and processing. Although at this stage, students will not understand many details of the interface design, they will gain an appreciation for its use.

Junior Year

The junior year carries the highest concentration of computer/chemometrics training. See Tables 1 and 2 for a summary of the relevant courses and goals. CMC Lab and Physical Chemistry are taken concurrently in the first semester. In CMC Lab students will refine their programming skills and develop their knowledge of important numerical and statistical methods. The current syllabus for this course is given in Figure 3. The course format consists of a lecture component which is strongly tied to a soft laboratory component. The lecture component primarily deals with the introduction and development of algorithms. This includes a justification of the algorithm's origin, its good and bad attributes, a comparison of its capabilities with related algorithms, a survey of its practical

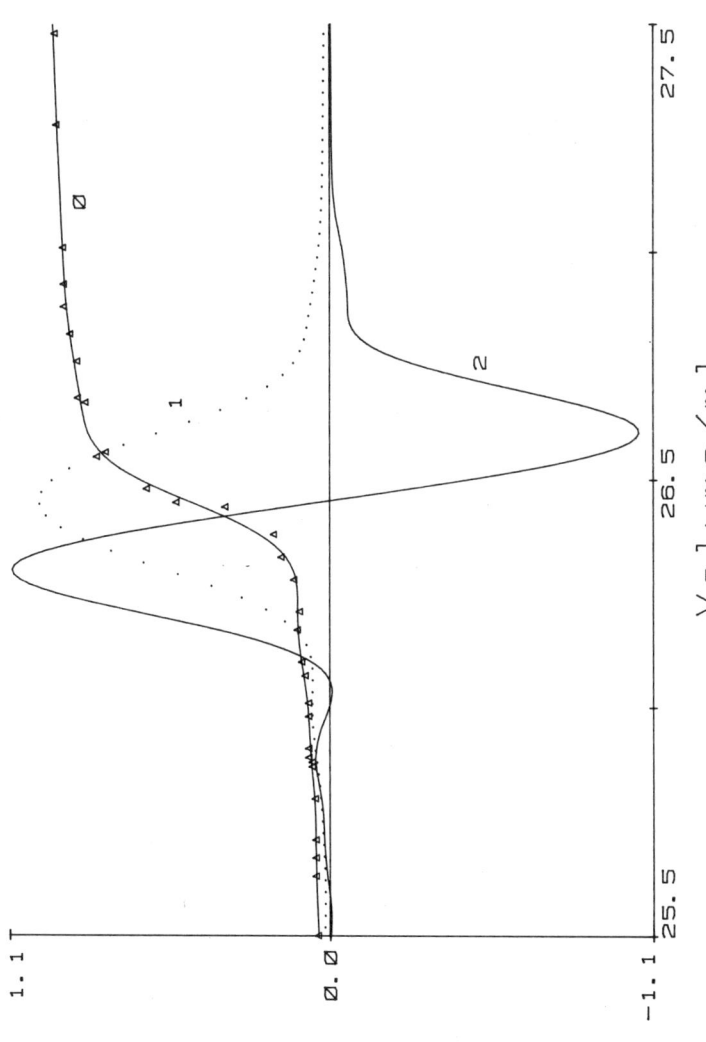

Figure 2.
Potentiometric titration data for the Fe(CN)$_6^{4-}$/Ce(IV) redox system. Ordinate scale has arbitrary normalization and units. Abscissa scale is titrant (0.05M Ce(IV) in 1N H$_2$SO$_4$) volume. Curve 0, raw($\triangle$) titration data (system potential as ordinate) and its fit superimposed. Curve 1, first derivative of curve 0 fit. Curve 2, first derivative of curve 1.

Computational Methods in Chemistry Lab   Syllabus

1. Computer Programming Refinement
   A. BASIC +
   B. System Utilities (PIP, Text Editor)
   C. FORTRAN Interpretation (as needed)
   D. Some Applications
         function evaluation
         graphics
         working with infinite series

2. Some Numerical Methods
   A. Quadrature
   B. Root Finding
   C. Interpolation
   D. Feature Location

3. Noise
   A. Distrubtions
   B. Central Limit Theorem
   C. Generators
   D. Simulation
   E. S/N Enhancement
   F. Error Propagation and Experimental Design

4. Regression
   A. L2 Estimation
   B. Matricies (some properties, some uses)
   C. Local Polynomial Regression, applied to
         smoothing, interpolation and differentiation
   D. Multiple Regression
   E. Nonlinear Regression

5. Convolution and the Linear System
   A. Integral Transforms
   B. Deconvolution, applied to
         transient kinetics and/or spectral resolution
         enhancement

6. Utility Program Concept
   A. Generalized Data Treatment
   B. Structured Programming

Figure 3.
Course syllabus for CMC Lab.

A College Curriculum Response to Chemometrics 31

applications in chemistry. In addition, programming tips, program structure concepts, and useful advanced features of the BASIC-PLUS language (under RSTS/E for the Digital Equipment Corporation PDP 11/70 CPU) are dispersed throughout the lecture phase of the course. In the laboratory component of the course (the computer and mathematical skills make up the laboratory apparatus), students are assigned projects which illustrate and/or extend the material presented in lecture. Many of the assigned projects carry a substantial programming aspect as well as require reference to source material not specifically detailed in lecture. The objective here is to teach students how to operate; i.e., how to deal with problems of a computer/chemometric nature. Students experience all aspects of a particular problem; the algorithm details, its application, strategy development, code development, implementation, and interpretation. Students are also required to maintain a detailed documentation notebook of all important software they write. This includes a brief description of the algorithm(s), its limitations, accuracy, etc., succinct strategy development, code details, parameter definitions, required subprograms, modification history, usage, brief user instructions, listings and sample menu. Proper documentation procedure is an important aspect of computer/chemometrics training. The students will discover this firsthand when they need to use this material in a later course.

A text which adequately covers a significant fraction of the topics listed in Figure 3 is not yet available. The current text for CMC Lab is the excellent applied statistics paperback "Data Reduction and Error Analysis for the Physical Sciences" by P. R. Bevington[2] which serves well for about one quarter of the course. Also, the BASIC-PLUS langage manual for the college RSTS/E operating system is strongly recommended. To fill the remaining gaps, a selected list of books is held on reserve in the college library for in-library use[3-9]. References to the original literature that would enrich a given topic or assist in the project work are provided by the instructor. Student performance is evaluated based on the project work, documentation notebook, written examinations, and a comprehensive final examination.

To illustrate the nature and level of topic coverage in CMC Lab, some aspects of the course syllabus (Figure 3) will now be discussed in more detail. Introductory remarks develop the concept of computer/chemometric enhancements in chemistry. How it all began with the theoretical chemist attempting to solve problems that did not have closed-form solutions and how this approach worked its way into the experimental laboratory. The impact of micro-computers on modern laboratory instrumentation is stressed as well as how these advances in instrumentation must be paralleled by similar advances in chemometrics if optimum information is to be derived from them. Several examples are given which illustrate the chemometric approach to problem solving. These include the role of simple simulation techniques in experimental design optimization, multiplex spectroscopy, and a detailed illustration of the use of simple chemometric techniques to resolve the infrared spectrum of a dilute component in a spectrally overlapping solvent matrix.

The first item of the syllabus requires about 2-1/2 - 3 weeks to get through. It appears that this induction period is necessary to bring students to a proper momentum level in the course. Computer programming refinement is not isolated here but, as noted earlier, it is dispersed throughout the course. The first few lectures do, however, concentrate on this subject by presenting a collection of concepts illustrated with examples. Some of these are: binary representable numbers, internal representation of data types, machine limitations, integer and floating point arithmetic, accumulation of round-off error, tips on the avoidance of serious round-off error, useful interpolation formulae, problem formulation, the algorithm, code development strategies, debugging strategies, efficiency concepts, structure concepts, fidelity evaluation. BASIC-PLUS is chosen as the programming language because it allows for the gradual development of programming sophistication, an important consideration in view of the aufbau structure of the computer/chemometrics teaching program. Students leaving this program will have learned computer programming and should be able to readily adapt to any programming language they might need by self-study. The ability to read and use FORTRAN is important in scientific programming applications since many useful algorithms have been coded as subroutines in this language. For this reason FORTRAN interpretation has been included here and is used throughout the course. Certain features of the BASIC-PLUS language allow for a virtually direct, one-to-one translation of FORTRAN into BASIC.

The applications listed under item 1 in Figure 3 serve to get students started in applying the programming and computational concepts noted above. For example, several function evaluation projects (probability distribution function shape comparisons, global titration curve generation), which produce high density data files for high resolution graphics display, quickly reveal loop arithmetic errors if the step increment is not exactly binary representable. The fundamentals of waveform and bit-map graphics programming (addressing and scaling algorithms) are introduced and projects are assigned to modify existing graphics software. This software was developed to emphasize structure, efficiency and, in general, good programming practice. In the projects, students are required to dissect and understand virtually all aspects of the code and in so doing learn a variety of programming techniques. This is an excellent vehicle for developing programming skills. The evaluation of series representations of functions with the computer is introduced. This application illustrates the concepts of approximation, iteration, convergence criteria, precision control, overflow, scaling and transformation. Series approximations are used in the evaluation of nonelementary integrals and their use in function evaluation where round-off error presents a serious limitation is illustrated. A project which applies these concepts, dealing with the virial representation of a Dieterici gas is assigned.

Item 2 of the syllabus (Figure 3) begins with the consideration of several numerical integration methods (histogram, Monte-Carlo, trapezoidal, Simpson, composite polynomial and Gauss quadrature). The theory of truncation and round-off errors as applies to quadrature is then developed. A project is assigned which produces a double logarithmic plot of relative quadrature error against interval number for three different quadrature methods, and vividly demonstrates the meaning of efficiency and round-off error associated with numerical integration. Results from a student project are shown in Figure 4. Finally, iterative quadrature with appropriate precision control is introduced and coded for future use.

In this section on numerical methods, several root finding algorithms and related methods are developed. These include: successive approximation, graphical solution, interval division, bisection, Newton-Raphson, convergence, x and y precision control, roots of unity, Cardan's method for cubics, Horner's method of polynomial evaluation. In lecture, only the location of single zeros is treated. Students extend this wrok in projects by developing algorithms for finding multiple zeros (both crossing and tangential) with simultaneous x and y precision control. Algorithms are tested with equations from real gas behavior and chemical equilibrium systems.

The next topic in this section treats generalized (unequally spaced domain) polynomial interpolation. Splines enhancement is also discussed. The application of these methods to essentially noise-free data structures is stressed. A project involving the transformation of a real us-vis absorption spectrum originally represented in an equally spaced wavelength domain to a representation in an equally spaced wavenumber domain is assigned.

The last topic in this section on numerical methods, feature location, actually overlaps into the next syllabus item. Simple iterative searching techniques for accurately finding peaks, valleys, and widths of data structures like spectra or chromatograms are treated here. A detailed project is assigned using the Plank radiation law as a model for a continuous probability density function. The project involves iterative quadrature with infinite integration limits, special analytical integration techniques, exposure to several important special functions, iterative searching, root finding and series approximation to determine the peak-location, peak-value, area, standard deviation, and half-width (fwhm) of the density function to five significant figure precision. Several independent methods are used so that the quality of the recovered parameters can be judged. This project, in addition to providing applications practice of previously learned material introduces new techniques and also sets the stage for an experimental design problem dealing with a "blackbody"

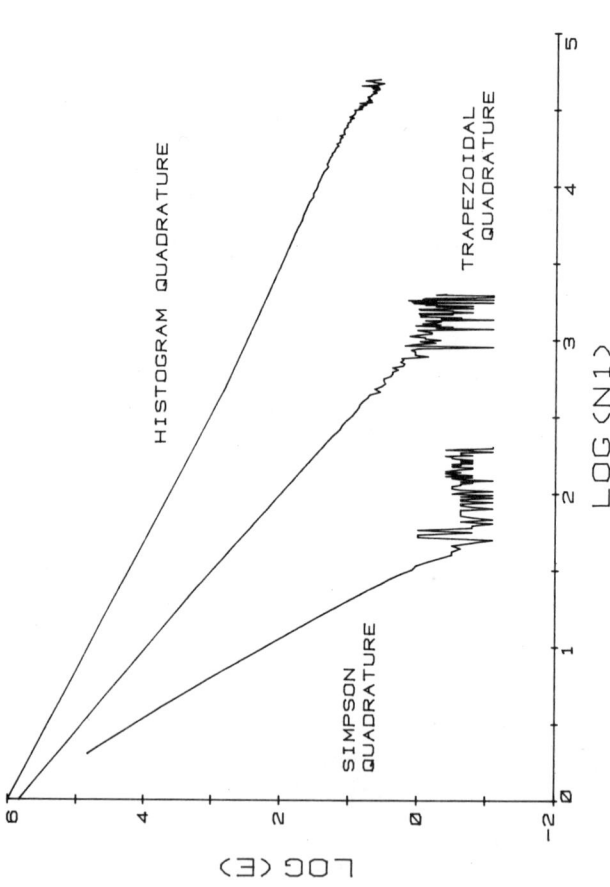

Figure 4.
Empirical functions representing the relative quadrature error (E) versus the number of subintervals (N1) for three different quadrature methods. The function integrated was $f(x) = e^{-x} - e^{-2x}$. E is defined as $|(Q-T)/T| \times 10^6$, where Q is the integral determined by quadrature and T is the true integral. The absolute value slope of each curve represents the efficiency of the corresponding quadrature method. The noise in the tail of each curve marks the onset of serious round-off error associated with the quadrature method and the calculation of E. The mantissa length used in these calculations was 23 bits.

A College Curriculum Response to Chemometrics   35

thermometer to come later.

In the third syllabus item (Figure 3), the properties (normalization, estimators, moments, dispersion relations) of discrete and continuous probability functions are introduced and compared. Both univariate and multivariate systems are considered. The role of the central limit theorem in justifying the normal error distribution for various types of experimental data as well as its use in the construction of an approximate Gaussian random number generator are covered. Other techniques of producing specific random number generators (eg. triangular, exponential, Poisson) are also considered[4]. A software multichannel analyzer (MCA) is developed to collect the generator distributions in pulse height analysis mode. The MCA has a variety of options which enhance its versatility such as variable ADC range, variable channel memory, optional multichannel scaling mode, statisitics option, collection status/control, source generator input via function subprograms and others. Hypothesis testing is introduced with emphasis on the χ^2 statistic, and is applied to evaluating the quality of source generators under various conditions of collection. One phase of the analysis demonstrates that event counting in each channel of the analyzer is governed by Poisson statistics irrespective of the nature of the source distribution. This concept marks the essence of an important instrumental detection technique (eg., photon counting). In cases where only a segment of the source distribution is collected, an interesting project evolves relevant to the χ^2 analysis of these data, which requires parameterization of the segmented model function. One approach involves iterative moment matching of the model distribution to the experimental distribution. These calculations require iterative quadrature to evaluate the moment integrals when the model function declares these as nonelementary (eg. a Gaussian model function). The results of these analyses are later compared with a direct, properly weighted (Poisson statistics) least squares fitting procedure. Another interesting aspect in this work concerns the detection of systematic error in the central limit theorem Gauss generator. The error is detected when collection results in a high signal-to-noise (S/N) ratio (eg. 500K counts in 100 channels) and when a substantial analog range is sampled (eg. ±3 standard deviation units). Under these conditions the noise level is sufficiently reduced to expose the systematic error which is primarily contributed by the tails of the distribution, a finding consistent with the central limit theorem approximation. This study simulates a real laboratory experience whereby systematic errors in the experimental apparatus are buried in the noise under conditions of low S/N but as the statistical significance of the measurement improves (eg. precision in the Gauss experiment above was increased by accumulating more total counts), the systematic errors in the apparatus begin to take over and become limiting unless reduced or eliminated. Also illustrated, is the fact that hypothesis testing methods in conjunction with careful control experiments can be used to expose systematic errors in a measurement system.

Another aspect of this section on noise introduces several real-time S/N enhancement techniques appropriate to both counting and time-sweep experiments. Simple post-processing digital filtering techniques are also treated. Techniques for adding various types of noise to data are illustrated and several simple simulation projects are assigned. In one such project a Monte-Carlo ensemble averaging experiment is developed to follow the S/N enhancement with scan number of a simulated spectrum starting with a S/N ratio of one and ending with a S/N ratio of about 300. The results of this project clearly demonstrate the square root advantage and disadvantage aspects of the approach. In another project, students simulate multipeak spectra with different levels of noise. Students are allowed to view a noisy spectrum once and retain minimal information about it, eg. number of feature peaks, approximate peak threshold level, and approximate peak height/width ranges. They are then asked to design an efficient algorithm that will automatically locate the main feature peaks of the noisy spectra by a combined filtering/searching technique. The project is completed with an accuracy evaluation of their algorithm (they know the answers, a plus feature of simulation work) and a critical appraisal of its limitations with example test cases that will mark its failure.

In the last topic covered in this section, several experimental design problems are discussed. One of these involves a gas-liquid chromatography peak-width "getter" and another deals with the "blackbody" thermometer noted earlier. In the latter experiment, several properties such as peak-height, peak-area, peak-location and peak-width of the blackbody radiation profile are considered as "temperature probes" for the thermometer design. Simulation techniques are used to optimize the performance of the thermometer, i.e., find the most sensitive probe. Finally, error propagation methods are introduced and are used to evaluate the fidelity of the experimental designs.

The fourth syllabus item in Figure 3 introduces least squares (L2) estimation from the point of view of maximum likelihood estimation based on a Gaussian parent population (although it is later pointed out that this is not the most general approach to formulation). L2 estimation is developed in lecture with a two-parameter linear model for the case of no weighting and includes estimation of parameter uncertainties. This derivation which is designed to illustrate important mathematical summation techniques and the use of the Kronecker delta, is extended in a project to include generalized weighting of the data. The concept of weighting the fit, methods to estimate weighting factors, goodness of fit criteria, and the consequences of metric transformation are treated in detail. A project involving the work-up of kinetics data is assigned which demonstrates the necessity

A College Curriculum Response to Chemometrics 37

of proper weighting with noisy data. (All of the project work on regression makes use of the previously learned simulation techniques to generate noisy data for testing the algorithms. This is a particularly appealing approach since the quality of parameter recovery must pass the ultimate test of truth.)

The properties of matrices and basic matrix algebra are considered next with emphasis on linear systems of equations and notation economy. Matrices are used extensively in the more general regression problems to follow. Next, polynomial least squares is introduced with the local polynomial regression problem, applied to smoothing, interpolation and differentiation of noisy data. The ill-conditioned nature of the polynomial problem is stressed. A marginal but highly structured version of a local polynomial curve fitting program which principally suffers from ineffective matrix inversion technique (the internal BASIC-PLUS matrix inverter is used directly) is given to the students. They are then assigned different projects to refine the code in a variety of ways which include accuracy flag and scaling enhancement of matrix inversion, iterative inversion, orthogonal polynomials, direct algebraic solution, efficient data I/O, storage and graphics enhancement.

The next topic develops the general multiple linear regression problem using matrix formalism throughout. Included in this development are generalized weighting, parameter uncertainty estimation and goodness of fit considerations. Student code is tested with the previously acquired MCA data. A pleasant surprise to some (those who performed correctly) is realized when the reduced χ^2 for the least squares fit matches closely with that obtained by definition in the earlier project. This section on regression terminates with the nonlinear least squares problem. Various parameter optimization strategies are presented and the topic is concluded with a detailed treatment of the Marquardt algorithm[10].

The superposition principle and other properties of linear model systems are presented in the next syllabus item of Figure 3. Laplace and Fourier transform methods are reviewed and the former is applied to a deconvolution problem involving transient kinetics data or the latter is applied to an idealized deconvolution problem dealing with spectral resolution enhancement using analytic models. The fast Fourier transform is reserved for the instrumental methods course where it will be illustrated in a simulation experiment involving multiplex ir spectroscopy and also applied as a digital filtering tool.

The final syllabus item in CMC Lab (Figure 3) introduces the <u>utility program</u> concept of processing data obtained from computer interfaced instrumentation. Each student contributes to the development of a command driven utility program which consists of a main program command director and a set of functional subroutine command responses that allow the user to perform various routine manipulations to process data already collected and stored on disk. Subroutines perform I/O to and from a divided utility program workspace and basic mathematical operations on data arrays brought into the work space. Typical operations include +, -, ÷, x two data block regions; scale, bias, $\int dx$, d/dx, $\log_a x$, a^x, smooth, interpolate, shift, min/max locate, window a selected data block region. The utility program is strongly coupled to a video graphics display by a DISPLAY command with overlay, find point, patch point, and xy expansion features for detailed visual examination of raw or processed data. A status routine is also included to document a specified processing cycle. A START status routine is called before you want to carry data through a chain of utility porgram processing operations which is then terminated with a STOP status command. The START command opens a file for the logging of all operations, parameters and results involved in the command chain up to the STOP command which closes the status file. A hard copy of the status file is readily obtained for your laboratory notebook.

The utility software, by definition, is not specific to a particular type of data but is of general applicability to data arrays obtained from any instrument output (several formating conventions have been adopted in the laboratory instrument interfaces). The usefulness of the utility package is quickly recognized by the students in the senior year instrumental laboratory. In this computer-based laboratory environment, <u>routine</u> manipulative operations are continually having to be performed on digitized data structures. The absence of utility-type software requires that a short, specific program be written each time you need to do this or that to your data. Not only does this become tedious but it quickly leads to a library of useless spaghetti-type software and, in addition, encourages poor organization, poor structure, poor documentation of results and, in general, bad habits. Newcomers to the computer-based laboratory must learn how to operate effectively in such an environment. This project, in addition to providing training in generalized data treatment concepts and structured programming technique, introduces the team design approach to problem solving, a common scientific practice. The development of the utility software involves considerable string manipulation and organizational detail.

Physical Chemistry, also a junior year course (Table 2), opens the door to numerous possible computer/chemometric applications. In fact, it has become traditional to teach some computational/computer methods in the Physical Chemistry course itself [11,12]. The existence of CMC Lab does not make this necessary. However, students are encouraged to <u>use</u> the computer/ chemometric methods they have learned, where they can be useful, in general problem solving throughout the course, eg. root finding to solve transcendental equations, quadrature for

A College Curriculum Response to Chemometrics 39

calculating thermodynamic properties from heat capacity data, graph preparation, Powell plots in kinetics, extrapolation applications, function feeling problems[13], etc. Recognize that this is not a computer-based studies approach[14]. The distinction is a clear one. In the former approach, students apply computer/chemometric methods in the subject when it is meaningful to do so, i.e., the methods are used as they would be used in a real chemical practice situation. In computer-based studies, students learn the subject by using carefully designed program packages. The purpose of computer-based studies in physical chemistry is to teach physical chemistry, not computer/chemometric enhancements.

The Physical Chemistry Laboratory taken in the second semester makes considerable use of instrumentation and data reduction techniques. Techniques learned in CMC Lab will be used extensively here. At present, two spectroscopic experiments in this laboratory utilize instrumentation that has already been interfaced to a laboratory computer for data logging. These deal with diatomic vibration-rotation spectroscopy[15], and the electronic spectrum of iodine[16]. Computer/chemometric techniques used in the treatment of data from these experiments include: accurate multipeak location (local regression/derivative techniques), general curve fitting and plotting, multiple linear regression, matrix techniques, position confidence intervals[17], and uncertainty propagation. Computer interfaces for experiments which cover other physical chemistry areas are currently under development. These include: a pressure transducer/thermister interface to directly recover Joule-Thomson coefficients, a bomb calorimeter/thermister interface to generate high density temperature/time profiles, a polarograph interface to study complex ion formation, and a mass spectrometer interface to acquire accurate appearance potential data. The emphasis associated with computer-enhanced experimentation in this laboratory is on the use of such experimentation. In addition, this use will be limited. Certainly, physical chemistry experiments can be performed without such enhancements. Although computer-assisted experimentation emphasis in Physical Chemistry Laboratory is not on interfacing, students with interests and abilities in this area will be encouraged to contribute to this aspect of the experimental design. An additional future objective of this laboratory is to provide students with a brief exposure to digital electronics and subsystems (gates, flip-flops, counters, A/D converters, D/A converters, transducers, operational amplifiers, etc.) that will serve as a foundation for instrument interfacing practice in the instrumental methods course. Several lectures, demonstrations and self-study assignments with some hands-on experience are anticipated.

Senior Year

In the senior year, three courses are involved in computer/chemometrics training (Table 2). The Qualitative Organic Analysis Laboratory is an ideal place to implement computer searching methods with a spectra library. This aspect of the course is currently under development and will include several lectures, reading assignments and assigned project work relevant to search algorithms, probability based matching, data base structure optimization, data base management, spectra condensation, prefiltering, etc. The objective here is to provide a degree of training in awareness, vocabulary, mechanics and practice in the area of information retrieval. Limited infrared and NMR spectra data bases will be acquired and used in this laboratory as an aid to unknown identification. The Chemistry Department's IR and NMR spectrometer/laboratory computer interfaces with a direct serial link to the college computer which provides data base creation/management software under RSTS/E will allow for routine access to the search facilities, when implemented. Also, on-line connection to a literature information service would be useful here. The College Library is currently tied in with the DIALOG system.

Computer/chemometrics applications in the Topics in Advanced Chemistry course will depend on the topics being covered. Currently, the course is equally divided into two isolated components, organic and physical chemistry. For the past several years the physical chemistry component has concentrated on theoretical chemistry and spectroscopy. Chemometric enhancements developed the eigenvalue problem in quantum mechanics which was applied to NMR spectra simulation[18,19]. With a different topic emphasis in this component of the course, other chemometric areas could be treated, eg., numerical solutions to differential equations in connection with photokinetics, polymer formation kinetics, or biomolecular matrix probe kinetics.

An important computer applications training objective in the Instrumental Methods course is computer interfacing design/practice with direct application to chemical instrumentation. Each student will be given one or two interfacing projects involving a standard I/O port and/or data domain conversion with event timing and control as appropriate. These projects are relatively simple and focus on the theme "How to do it" for standard laboratory applications involving standard laboratory instrumentation, eg., data logging of a digital or analog signal, ensemble averaging, real-time x/y drift correction, real-time boxcar averaging, real-time signal ratioing. Students work out the hardware/software details themselves. With a good general purpose hardware/software support medium available, this will only require students to lay out the basic logic design, timing, and signal level amplification. The Chemistry Department's user-friendly Digital Equipment

A College Curriculum Response to Chemometrics 41

Corporation MINC laboratory computer provides this kind of support which will enable project completion in a reasonable period of time.

Emphasis in these projects is on interface design in an environment with "off-the-shelf" modular hardware/software components[20]. There is simply not enough time in this instrumental methods course to treat areas like assembly language programming and the design/construction (electronics) details of interfacing hardware. Thus, interfacing projects are kept simple, stressing routine laboratory applications. These projects are designed and tested so that they can be completed in one or two weeks. More advanced work can be given to students so motivated in a mid-term projects course. Several orientation lectures on basic interfacing principles and the MINC system precede the interfacing project work. This, together with MINC BASIC, excellent MINC documentation, and individual help sessions with the instructor, carries students through their projects.

In addition to interfacing projects, students in this course carry out a number of instrumentally based experiments of a quantitative analytical or physical nature, with or without the aid of computer enhancement, as appropriate, always paying proper attention to data reduction details. In the computer-enhanced experiments, students use instrumentation that has already been interfaced to the laboratory computer for <u>standard operation</u> which means data logging often with signal averaging capability. The only computer/instrument I/O capability that was available in the first year's implementation of this project was RS232C bit-serial. This allowed for standard operation interfacing of a UV-VIS spectrometer, an IR spectrometer, a fluorescence spectrometer and a hard copy digital plotter (for general purpose plotting and semi-automatic digitizing) to the laboratory computer. In addition, a communications link was established between the lab computer and the college computer.

A variety of interfacing and chemistry experiments were designed around this available equipment. Each student was assigned one interfacing project which involved one or more of the following: data logging, ensemble averaging data acquisition, boxcar averaging data acquisition, instrument control, handshaking, data format conversion, real-time graphic display. A typical experiment description for one of these projects is given in Figure 5. The chemistry experiments that were computer-enhanced by means of data logging and, in some cases, by signal averaging and, in all cases, by chemometric post-processing are listed below. (All quantitative experiments in this laboratory carry some degree of chemometric enhancement regardless of how the data were obtained.)

Chem 410A Experiment

Title/ Bit-serial interfacing, ir data logging and MINC enhanced peak picking

Major Instrumentation/ PE 283 ir spectrometer
/ DEC MINC laboratory computer

Primary References/

    (1) 283 ir instrument manual
    (2) MINC BASIC documentation
        /Book 6; CIN, COUT
        /Book 7; Chapter 7
        /Book 4; GRAPH, 'MOVE', FIND-POINT, INDEX ARRAYS

Minimum Expectations and Guidelines/

Design and code a computer program that will allow spectral data logging, sequential file storage*, and graphic display (with MINC enhanced feature location) from the 283 ir spectrometer.

1. Determine RS232C pin configuration standard for both computer and instrument. If in doubt use schematics or oscilloscope. Confirm assignment with instructor before hardwire connection.

2. Confirm approach and design with instructor before implementation.

3. Acquire a polystyrene film spectrum (ca. 2000 data points); store in a file; your collection program should include the 'strip chart mode' option of graphic display if the number of data points collected is in excess of 512 points.

4. In a separate program use the FIND-POINT routine and INDEX option of GRAPH to peak process several features in a 512 element segment of your collected data. Compare both methods and comment.

*See instructor for data file formating protocol.

Figure 5.
Interfacing experiment prescription in Instrumental Methods Lab.

A College Curriculum Response to Chemometrics 43

(1) Förster Cycle Determination of an Excited State pK
(2) Spectral Detection Limit of a Fluorescence Spectrometer
(3) Fluorometric Selectivity, the Spectral Resolution of a Binary Mixture
(4) Performance Parameters of a Spectrometer
(5) Radiative Lifetimes from Electronic Spectra
(6) A Stoichiometry Study of a Metal-Ligand Complex
(7) Multicomponent IR Analysis
(8) Dilute Component IR Analysis

Essentially all aspects of the CMC Lab course were illustrated in these experiments. Students were required to choose one experiment in each instrumental area (IR, UV-VIS, or fluorescence spectroscopy). This policy also applies to other experiments in this course. A typical chemistry experiment prescription, illustrating the first experiment in the above list, is given in Figure 6. It is clear from the experiment "description" in this figure that in this senior level laboratory experience, students are expected to bring together a number of resources relevant to the design and implementation of their experimental work. Also, lectures on the various instrumental methods cannot be kept in phase with the experimental work since, due to the limited number of instruments, these instruments must all be in use from the beginning of the course.

Even with the limited computer interfacing capability available in this first year implementation, a sense of excitement pervaded the laboratory atmosphere as students experienced state-of-the-art technology and modern experimental design concepts applied to solving chemical problems. Utility programs and chemometric techniques were used routinely in this laboratory. (At this stage of program implementation, students were prepared in computer/chemometric methods either in the first semester CMC Lab or in a similar mid-term course taken two years previously.) Computer/chemometric techniques were viewed as a natural tool to problem solving in the chemical laboratory. The Chemistry Department has subsequently acquired a variety of general purpose I/O modules for the laboratory computer which are currently being incorporated into the course. These will provide interfacing capability for instrumentation not equipped with standard digital I/O ports.

A few comments about the laboratory computer/instrument environment are appropriate. The MINC laboratory computer system is currently configured with a PDP 11/23 CPU, 256 K bytes of addressable memory, a dual floppy disk drive (0.5 M bytes each), four RS232C ports, an IEEE-488 bus interface, a 180 character/sec line printer, and a package of general purpose I/O modules. The later items include a 16-channel A/D converter (12-bit resolution), a programmable clock with two Schmitt triggers, a 4-channel programmable preamplifier, a 16-channel digital input unit, and a 16-channel digital output unit.

Chem 410A Experiment

Title/ Förster Cycle Determination of an Excited State pK

Major Instrumentation/ Beckman DU-8 uv-vis spectrometer
/ SLM fluorescence spectrometer
/ MINC laboratory computer

Primary References/

(1) Z.R. Grabowski and A. Grabowska, Z. Physik. Chem. Neue Folge, **101**, 197(1976).
(2) A. Weller, in "Progress in Reaction Kinetics," Vol. 1., ed., G.N. Porter, Pargamon, Oxford, 1961, p. 189.
(3) C. A. Parker, "Photoluminescence of Solutions", Elsevier, New York, 1968 (LND Reserve).

Minimum Expectations and Guidelines/

The equilibrium constant for the proton transfer reaction $AH^* + H_2O \rightleftarrows A^{-*} + H_3O^+$, where AH^* denotes 1st singlet excited 2-napthol, will be determined via Förster Cycle methodology. This experiment illustrates a nice connection between electronic spectroscopy and thermodynamics.

1. Use pH ~ 0 to force AH dominance and pH ~ 13 to force A^- dominance.

2. Fluorescence spectra require correction to quantum flat detection.

3. Initial aqueous stock solutions of AH and A^- should be about ~10^{-4}; use concentrations ~10^{-5}M in fluorescence measurements to avoid inner filter effects.

4. All spectra will require mathematical processing, so be sure to log data.

5. Determine $\bar{\nu}_{oo}^J$ by the three methods outlined in (1), and compare results.

*Of course, meticulous attention to cleanliness detail is required.

*Fluorescence spectra (F) in an energy $(\bar{\nu})$ and a wavelength (λ) representation are related as follows: $F(\bar{\nu}) \, d\bar{\nu} = F(\lambda) \, d\lambda$.

Figure 6.
Chemistry experiment prescription in Instrumental Methods Lab.

In addition, medium resolution graphics is provided on the MINC video terminal and hard copy high resolution graphics is available on a Hewlett-Packard 7225A digital plotter. Any instrument output brought into the computer can be displayed on these devices.

The computer is located in the same laboratory which contains the major instrumentation that will be interfaced to it. The MINC system operates in a single user, real-time capacity under MINC BASIC, which is a combined operating system and programming language (a separate RT-11 operating system which provides access to FORTRAN and assembly code is available). In the standard operation environment noted earlier, this implies that only one experiment can be tied into the computer at a time. The RS232C instruments are switch selectable from a panel located on the computer chassis. Instruments requiring the general purpose I/O modules are connected to them by easy access plug-in of edge connectors coded to the instrument. MINC BASIC can communicate with all the I/O hardware in the system.

The MINC/college computer (11/70) communications link noted earlier serves an important function in this laboratory by allowing the single dedicated laboratory computer to extend into a multi-user configuration for pre- or post-processing applications. If the MINC is going to be busy with an experiment, previously acquired data can be transferred to the 11/70 for processing or MINC software can be developed with the 11/70 text editor and downloaded to the MINC when the latter becomes available.

An NMR spectrometer, a mass spectrometer, a gas chromatograph, and a polarograph will also be included in the list of standard operation interfaces. This laboratory facility is available for other chemistry courses and for general departmental use.

A final computer/chemometric enhancement aspect in this course involves the development of simulation experiments to illustrate areas where instrumentation is not available at Loyola such as ESCA, nuclear activation analysis, fast kinetics or Fourier transfer spectroscopy. Software will be written to simulate actual instrument operation from a keyboard, leading to the final output and display of noisy data. Students will be asked to "run" the experiment, collect the data and analyze these data by methods which they will choose and assemble.

ACKNOWLEDGEMENTS

The work described in this article was supported in part by a grant from the National Science Foundation (SER-8162635). The author wishes to thank Edward Ferrell for providing the graphics software and some of the data used in producing several figures reported here.

REFERENCES

1. I. E. Frank and B. R. Kowalski, Anal. Chem., $\underline{54}$, 232R (1982).
2. P. R. Bevington, "Data Reduction and Error Analysis for the Physical Sciences", McGraw-Hill, N.Y., 1969.
3. W. S. Dorn and D. D. McCracken, "Numerical Methods with Fortran IV Case Studies", Wiley, N.Y., 1972.
4. R. W. Hamming, "Numerical Methods for Scientists and Engineers", McGraw-Hill, N.Y., 1973.
5. T. R. Dickson, "The Computer in Chemistry", Freeman, San Francisco, 1968.
6. K. J. Johnson, "Numerical Methods in Chemistry", Dekker, N.Y., 1980.
7. A. C. Norris, "Computational Chemistry", Wiley, N.Y., 1981.
8. D. L. Massart, A. Dijkstra and L. Kaufman, "Evaluation and Optimization of Laboratory Methods and Procedures. A Survey of Statistical and Mathematical Techniques", Elsevier, Amsterdam, 1978.
9. W. C. Hamilton, "Statistics in Physical Science", Ronald Press, N.Y., 1964.
10. D. W. Marquardt, J. Soc. Ind. Appl. Math., $\underline{11}$, 431 (1963).
11. R. R. Roskos, "Problem Solving in Physical Chemistry", West, N.Y., 1975.
12. D. L. Peterson and M. E. Fuller, J. Chem. Educ., $\underline{48}$, 314 (1971).
13. N. C. Craig, D. D. Sherertz, T. S. Carlton, M. N. Ackermann, J. Chem. Educ., $\underline{48}$, 310 (1971).
14. G. M. Barrow, J. Chem. Educ., $\underline{57}$, 697 (1980).
15. D. A. Aikens, R. A. Baily, G. G. Giachino, J. A. Moore and R. P. Tomkins, "Integrated Experimental Chemistry", Vol. 2, Allyn and Bacon, Boston, 1978, p. 681.
16. I. J. McNaught, J. Chem. Educ., $\underline{57}$, 101 (1980).
17. E. Heilbronner, J. Chem. Educ., $\underline{56}$, 240 (1979).
18. J. S. Sims and G. E. Ewing, J. Chem. Educ., $\underline{56}$, 546 (1979).
19. H. Bauer and K. Roth, J. Chem. Educ., $\underline{57}$, 422 (1980).
20. S. P. Perone and J. F. Eagleston, J. Chem. Educ., $\underline{48}$, 317 (1971).

MICHAEL F. DELANEY AND ALPHONSE J. ANTONITIS

CHAPTER 3

Information Processing in Chemistry

1. Introduction.

Information processing has become an integral part of research. The initial stage of a research project consists of searching existing knowledge bases (library books and journals) to see if the desired information is already known and available. In the middle section of a research project, experiments are designed and conducted to acquire the needed information. In the final part of a research effort, the results are put into a form suitable for incorporation into the general pool of knowledge.

In a similar fashion, the more routine laboratory activities of a physical scientist involve a large quantity of information processing. Planned activities for a laboratory must be communicated to the laboratory staff in sufficient detail to insure that the tasks are accomplished in a manner which is efficient and accurate. The laboratory workers need to take advantage of existing data and procedures so that precious, and expensive time will not be wasted redeveloping known methods. Final results must be drawn out or derived from the raw experimental data, and used to generate the information in the form needed.

The means for incorporating information processing into scientific research already exist and are continually being improved. The dedication and imagination of many scientists have resulted in a multitude of applications of this most useful field. New tools for information processing are constantly being created and many existing methods are being updated for use in scientific research. More scientists should become comfortable with taking advantage of enhanced information processing tools as they become available. In fact, researchers should become actively involved with the development and implementation of these capabilities.

In the following section we will discuss a general philosophy of computerization, which we embrace. We will then present a few of the topics from analytical chemistry that especially lend themselves to the incorporation of computerized information processing techniques. We have used each of these areas, to a lesser or greater extent, in both undergraduate and graduate chemical education.

1.1 Scientists need a thorough background in computerized information processing.

Our basic conviction is that scientists need a thorough background in computerized information processing. They should learn to 'do' computers, or else computers will be done to them.

Historically the development of computerized chemical information processing, and the accompanying learning activities, have been a series of dramatic advancements coupled with painful atrocities. This is a result of the actions of those who strived to make a useful field of information processing along with those who allowed themselves to be intimidated. A good amount of perseverance is required when one first starts to joust with a computer. Hard fought early battles will be rewarded with increased efficiency and time savings.

Past Atrocities.

Our goals for rapid, efficient, and effective chemical information processing must be coupled with an eye towards maintaining our sanity and our professional stature as scientists. We do not want to become computer engineers, but why should they have all the fun. This could be referred to as 'The Humanization of Computers'. The early application of computers in chemical research and education, as in other fields, could be referred to as 'The Computerization of Humans'.

Many of us might remember some of the early developments in scientific applications of computerized information processing. For some these past atrocities might have a high degree of nightmare quality. Programmers would live in fear that someone might shuffle their decks of IBM cards. And why did card punches refuse to print the contents of the card on top? Were the card punches embarrassed by what they had generated?

A quantum leap forward was made by the availability of interactive computing using an installation's main-frame system. In this configuration a much larger number of users could get frustrated simultaneously. We can remember a situation in which a new graduate student taking an instrumental analysis laboratory course dutifully collected 10 measurements for each standard solution and carted these results over to the computer center. He carefully entered his dozens of data points into a canned program, working his two fingers to the bone. What did he get in return (bony fingers), an impolite 'INSUFFICIENT STORAGE', whereupon his data was promptly flushed by the system, so it could go on and perform an unspeakable act on the next user. This student swore off any further use of computers in his graduate studies. As it turns out, he got a job with a large computer-instrument manufacturer, and now announces that he has 'gone digital'.

The next development was the use of on-line laboratory minicomputers. In this circumstance, the user interface often consisted of unforgiving format specifications for input. This meant that the unsuspecting person would probably either crash the computer or not actually be entering the desired numbers. The experienced user was obliged to be subjected to the dehumanizing activity of counting columns. This was also the era of the 110 baud s-l-o-w teletype printers.

We have now progressed to the point where students are provided with fast and elegant video display terminals in the computer center. Unfortunately, they either have to have one willed to them by a student that has expired, or pay someone to eliminate a fellow student. At times, the computer center is akin to Filene's Bargain Basement on a sale day.

We feel that the processing crunch and shortage of terminals which most academic computing centers experience is one of the major obstacles to educating science students in the use of computers. The average science student does not have the time to hang around the center waiting for a terminal to become available so that he can hone his skills. There is always a new set of organic reactions which need to be studied or some physical chemistry problems to solve.

Science departments should integrate computers into their courses. Students could then acquire valuable computer skills while studying in the context of their chosen scientific field. Any science education would certainly benefit from the use of a computer. These integrated skills would strengthen and further the learning process in a manner which is stimulating and enjoyable.

1.2 Modern instrumentation and analytical techniques require computers.

Today's analytical instruments are capable of acquiring data faster and to a higher degree of precision than ever before. Coupled with advances in technique, a modern analysis will often yield more data than a human can interpret. In many cases it becomes necessary to use a computer to take full advantage of the information generated by an experiment. The computer can be integrated into the instrument itself, interfaced to perform a variety of functions, or may be resident off-line. In any case the most important part of the instrument-computer interface is the scientist. The scientist must have the knowledge to make intelligent choices regarding computer hardware and software. He must be able to program in a structured manner and should understand basic programming principles. Without some background in computers, the full value that computerization offers cannot be realized. The key ingredient, as in any field of science, is the scientist himself.

It is a rare instrument today which does not come equipped with some type of computer implemented in its design. Microprocessors have come to replace the mechanical, often temperamental, subassemblies of many common laboratory instruments. In most, if not all cases, the role of the computer is to increase instrument throughput, improve precision of data, and help make the analysis easier to perform. The scientist is faced with a problem which these sophisticated machines induce. What does one do with all the data? Once an analysis becomes automated and hence easier to perform, it is usually performed more often. A computerized instrument can take a great deal more data points than a chart recorder, or a graduate student with a notebook. Unfortunately much of this increase in information goes unused. The user gladly accepts the increase in precision, is tickled with the ease of operation, but then does little or nothing with most of the information provided by the instrument. The step after analysis should be to squeeze every last bit of useful information out of the copious amounts of data generated by the machine. Computerized information processing is a means of optimizing the data we receive from modern instrumentation.

Information processing consists of a variety of software and hardware tools designed to aid the scientist in his research. It is the practice of using computers to help automate and optimize experimentation. Applications of information processing are as varied as the research which it is designed to support. Computers can be used for experimental control as well as for data acquisition and analysis. Instruments without microprocessor control can be interfaced to a laboratory minicomputer. Data can be acquired by the same minicomputer and analyzed or loaded onto a larger computer for interpretation. The limiting factor is the scientist's own knowledge of information processing.

Scientists should learn the applications and tools of information processing early in their careers so that the benefits of these techniques can be applied in future study. The use of these tools will result in a greater understanding of the current computerized methods and will aid in the scientist's own development of useful information processing tools. One of the more important aspects of this field is the ability of the scientist to create his own techniques and programs. Many packages are commercially available but these are only a small representative of the tools which could be available if more scientists integrated computers into their work.

Information Processing in Chemistry 51

1.3 Information processing training should be integrated into the science curriculum.

Information processing skills are too precious to be partitioned into specialized elective course which are too easy to avoid. The development of these techniques should pervade the traditional curriculum.

1.3.1 Computer fluency is essential.

The integration of computerized information processing techniques and strategies appears to be advisable in training scientists for their future careers. In addition to this exposure to information management, being able to program computers effectively and efficiently seems highly desirable. The dual concepts of structured programming and data structures should be stressed to instill in scientists the logic and order that will serve them well as they develop sophistication in the manipulation of data and derived results.

1.3.2 Students should be provided with professional tools.

Since computerized processing is going to greatly change the way scientists handle and assimilate information, the philosophical and instructional aspects should be 'hidden in a very visible way' in the educational curriculum. Rather than set aside specific courses, or portions of courses, the use of computers should be allowed/obliged all the way from General Chemistry 100 up to Doctoral Thesis 999. The scientist should have the same familiarity with 'his' computer as he does with his hand-held calculator today or his slide-rule fifteen years ago.

One very important aspect of this integration is to provide students with professional quality software and hardware tools. We loosely define a tool as anything which is useful. Some of the capabilities that are most important are the following:

o   A Fast Processor - The primary activities of accessing data, performing calculations, and examining results should be sufficiently fast as to be limited by the speed of the student's comprehension of what is happening, and what needs to be done next.

o   Adequate Storage Space - Ample resources for data, programs, and results will foster creativity and encourage students to be explorational. Students should learn the magnitude of their resources, and the management of space at an early stage. They should become familiar with the tradeoff between storage and regeneration of information.

o   Screen Editors - An important tool for obtaining high productivity is an efficient screen-based editor. In addition to cutting down on the need for paper, and providing a higher degree of reliability than printing terminals, screen editors are much more effective at keeping up with the frenzied pace of the human mind.

o   Symbolic Debuggers - Another surprisingly serviceable service is the symbolic debugger. We all make mistakes, and we should all expect to find them in a reasonable amount of time, using a logical procedure. A debugger allows the user to zero in on the offending block of code, and step through the execution process.

- Liberal Access to Terminals – The 'Secret World of Computing' stems originally from the fact that large computers needed to be kept securely locked away in clean, climate controlled rooms. In addition, computer systems use passwords, access protection, and privileges. Today it is common to find mini- and superminicomputers in the 'not so pristine' environment of the average research laboratory.

- Professional Quality Graphics – A picture is, of course, worth 2000 bytes (actually 4000 bytes on our VAX), but the generation of a graphical image should not be an involved or objectionable task. High quality graphics software and hardware devices should be liberally provided. As we will demonstrate later, the software should be able to handle a wide range of devices, since this independence will aid the users in focusing on an analysis of their information, rather than on the subtleties of each hardcopy device.

- Sophisticated Text Processing – We expect our students to be able to clearly express themselves using the written language. We want programs to be carefully and completely documented. We want written material that is presented to us to be readily legible. Each of these desires are aided by providing students with professional quality text processing facilities. This can presently be accomplished on any respectable computer system. At Boston University, Chemistry Professor Dan Dill has developed an integrated text processing system (TXTSYS) which runs on DEC computers, sending output to any type of printer or terminal. Using a specific type of hardcopy printer (NEC Sprinwriter), this software facilitates the use of super- and sub-scripts, special and mathematical symbols, and formatting equations, such as this:

$$D_{Lm\gamma}^{(-)v_f v_i} = \langle v_f | D_{Lm\gamma}^{(-)} | v_i \rangle = \int dR \, X_f(R) \, D_{Lm\gamma}^{(-)} \, X_i(R) \, . \qquad [1]$$

TXTSYS uses 'carat' (↑) to indicate super-scripts (up-arrow), 'bar' (|) to indicate sub-scripts, and backslash (\) to indicate that the next character is a special symbol. For example, backslash-m, gets translated in to the greek letter µ. This simple to use convention is processed into the nearly undecipherable escape sequences used to position the print thimble and carriage of the Spinwriter.

The use of TXTSYS has greatly aided technical manuscript preparation in our department (including this chapter).

1.3.3 Present computerized applications should be studied.

The programming instruction and tools described above would allow students to pursue virtually any kind of information processing and management activity or investigation imaginable. Science students should then be in a position to develop ideas of and on their own. These might include new ways of extracting results from experimental data or approaches to systematizing and organizing a knowledge base.

The alternative to having students program themselves to death is to study existing computerized scientific applications. This shows the relevance of information manipulation in solving real problems. Two of the chemical examples will be briefly mentioned.

Information Processing in Chemistry

o   Computerized Instrumentation.

   Computerized laboratory equipment is becoming increasingly common. Configurations consisting of computers integrated into the design of an instrument or attached and interfaced to an instrument are seen. Either approach should be able to serve as a testing ground for evaluating control and data acquisition strategies. The study of computerized instruments is especially appropriate in cases where the system was developed in the local laboratory, as this will mean that documentation will be available to specify the detailed operation of the hardware and software.

o   Bibliographic Search and On-line Data Compilations.

   One of the more useful advantages of computerized instrumentation and its associated data acquisition is the benefit of on-line data compilations. Data can be analyzed as it is acquired by the instrument thus allowing the user to make adjustments in the manner in which the analysis is progressing. In addition to this is the savings in time which is realized by interpreting the data as it is being acquired. This keeps us even with the frenzied pace at which our modern instruments pump out data.

   Particular applications which take great advantage of these capabilities are infrared and mass spectral analyses. Large libraries of known compounds exist for these types of analysis. With library searching such an important part of interpretation in these two fields, it has become quite important to devise methods to optimize the effort required to search through a library. Along with accuracy, an important factor in library searching is speed. Clearly a savings in time is realized if the searching is done while additional compounds are being run. Most manufacturers of these instruments now offer some type of on-line search systems.

   On-line search of literature compilations (Chemical Abstracts, etc.) are becoming so pervasive that an early exposure to such techniques in chemistry courses would seem to be mandatory.

1.4   Structured Information and New Programming Languages.

Once the data is gathered and a method of analysis is chosen, the chemist needs a means of conveying this information to the computer. At his disposal are a variety of structured programming languages, some of which are more suited to a given problem than others. The important realization is that there is no one programming language which can best serve all the needs of a laboratory environment. The scientist must adapt his choice of programming language to the application at hand. A knowledge of what languages are available and the best ways to implement them is essential in a modern laboratory environment.

In the past it seems that scientists were taught (or picked up off the street) FORTRAN with the understanding that this was the language for the laboratory. FORTRAN is still the most popular scientific language, almost in defiance to the many structured languages that have been developed since its inception. The newest version of the language, FORTRAN-77, is quite a bit more civilized than its predecessors but still is not all that the modern chemist should consider adequate. FORTRAN-77 is a good general purpose language which is easy to program and useful in environments where speed is not essential. The many improvements in the current version, such as IF-THEN-ELSE control, allow programming in a structured manner. The language is useful for manipulating data and performing calculations. This new structured version is an adequate language for introducing computer programming into an undergraduate science curriculum.

One drawback to a high-level language such as FORTRAN is speed. Ease of programming and user 'friendliness' are time-costly features. High-level languages are, in most cases, poor choices for real-time instrument interface. Instrument interface requires software, and hardware, which can control the experiment and acquire data at the speed which the analysis is progressing. For most laboratory applications, a mini- or microcomputer is used. The fastest control and data acquisition is realized when the software is written in the machine, or assembly language of the computer. For most chemists, programming in an assembly language seems a chore which is best left up to computer programmers. To get the most utility and fastest throughput from one's computer and instrumentation though, assembly language is a must. High-level languages lack the speed which is required for control at the instrument-computer interface. There is some consolation, though. High-level languages such as FORTRAN can call assembly language subroutines which can perform the time critical steps and then return to the 'user-friendly' environment of the calling program. A scientist who is 'up to speed' in a computer sense should be familiar with the assembly language of his laboratory machine.

One language which helps bridge the gap between assembly and high-level languages is C. Written at Bell Laboratories, C is a structured, high-level language which is capable of accessing hardware directly (1). This makes it useful for data acquisition and instrument control. It is not quite as alien to users of high-level languages as is an assembly language. Many users have found that programs written in C are easier to debug than in other languages and code written by other programmers is easy to follow. This is a direct result of the structure of the language and the style expounded by its creators.

Many programming languages have been developed for specialized functions which can readily be adapted to scientific applications. To best solve a given problem, it is important to choose a programming language which is suited to that problem. For example one should use an assembly language for instrument interface and a high-level language of some kind when working with structured data, or performing calculations. One language which offers such potential specialized utility is LISP (2).

LISP's primary use is in manipulation of non-numeric data. In computer science and information processing it is used extensively in artificial intelligence applications. We can see its use as a programming tool in spectral structure elucidation as it is ideal for programming decision making rules for interpreting structure. LISP's forte is its ability to allow recursive program structures (see section 1.4 of this chapter for an example of a recursive reference). It can also be used interactively which makes it quite useful in a dynamic laboratory environment.

A compact comparison of the popular commercial languages, with examples, has recently appeared (3).

Several concerted attempts have been made to create programming languages exclusively for the interpretation of spectra and manipulation of chemical structures. CONCISE (Computer Oriented Notation Concerning Infrared Spectral Evaluation), a language developed by Hugh Woodruff and Graham Smith, (4) (Merck Sharp and Dohme Research Laboratories, Rahway, New Jersey) is used to write an infrared spectral interpreting program called PAIRS (Program for the Analysis of IR Spectra). This English-like language allows the chemist to readily alter the rules which the program uses to base its decisions. The following is an example of some CONCISE source code from the program PAIRS:

<u>CONCISE</u> <u>RULES</u> <u>FOR</u> <u>C-H</u> <u>STRETCHING</u> <u>REGION</u> <u>FOR</u> <u>THE</u> <u>AROMATIC</u> <u>CLASS</u>

```
IF ANY INTENSITY 1 TO 10 SHARP TO BROAD PEAKS ARE IN RANGE 3050 TO 2990

    THEN BEGIN
        SET AROMATIC TO 0.20
    DONE
    ELSE BEGIN
        IF ANY INTENSITY 1 TO 10 SHARP TO BROAD PEAKS ARE IN RANGE
            3150 TO 2990
            THEN BEGIN
                SET AROMATIC TO 0.10
            DONE
            ELSE BEGIN
                IF SPECTRUM RUN IN OIL
                    THEN BEGIN
                        SET AROMATIC TO 0.05
                    DONE
            DONE
    DONE
```

As can be seen, the different weight factors can be easily changed and provisions can be programmed to allow for different experimental conditions.

Other language developments of specific chemical application have been in the field of organic synthesis. At Harvard University, a group under the direction of E. J. Corey (5,6) has developed a program for organic synthesis. LHASA (Logic and Heuristics Applied to Synthetic Analysis), is an interactive program which deduces retrosynthetic routes to find possible precursors to a target molecule. All of the information pertaining to chemical reactions in LHASA is programmed in an 'English-like' language called CHMTRN (Chemistry Translator). This interpreter takes information about reactions and structure as input by the chemist and sends it along to the main program (written in FORTRAN). This allows new data to be incorporated using a format which is easy for the chemist to understand and use. Many other synthetic and interpretive programs exist, most with equally clever acronyms (e.g. SECS), and the reader is urged to read further about these exceedingly useful tools (7).

As we have tried to show, although through a survey which is by no means comprehensive, scientists must learn to take full advantage of the computational tools which have been developed. Although the primary applications may be far removed from our own requirements, we should take the effort to survey what is available and adapt it to our work. A structured background in the most efficient methods of programming should be part of every science education. The time and effort saved later in the career, or even graduate training of an individual, would be enhanced greatly by the increase in productivity which could be realized by such knowledge.

1.5 Graphical Display of Information.

Scientific investigations are said to be conducted using a method in which hypotheses are formulated, tested, and evaluated. This process is repeated until a 'true and correct' view of the world is achieved. For most of us, at least the day-to-day activities, if not the whole of scientific research, is a series of coincidences, fortuitous accidents, and novel insights. We occasionally make unbelievable progress, and at other times, none at all.

The fantastic progress we are able to make comes about by a combination of thought and observation. Our thought process is fueled by experimental results, theoretical relationships, and the correlations which relate data and theory. The progression from hypothesis to proof and understanding is most certainly governed by the manner in which we interpret our results.

The assimilation of information is assisted by the preparation of graphical presentations of data, in the form of graphs, bargraphs, histograms, etc. The ability to take mass quantities of information, digest it, and incorporate the essence into a cogent view of the universe, should not be limited by the tedium of progressing from a table of numerical values and related equations to a graphical representation, from which our conclusions can be drawn.

We have trained a laboratory slave to sit by our instruments and to patiently record the results of our experiments. This slave is prepared to work for us around the clock, including weekends. This slave does not take coffee breaks, does not talk back to us (at least not verbally), and doesn't even expect a kind word in return. No, this slave is not a graduate student it is a DIGITAL COMPUTER.

Our next realization is that the laboratory slave is ready, willing and able to plot our graphs for us. We just have to tell it how to transform tables of numbers into pretty pictures. But are we ready to sit down with our laboratory computer and discuss with it the intricacies of autoscaling, multi-dimensional projection, polynomial smoothing, non-linear regression, and plot device driving? Probably not. Fortunately, someone has probably taught the twin brother of your laboratory computer to talk intelligently to the particular plotting device that you have available. Software packages for the graphic representation of data exist for a wide variety of computer processors and related hardcopy peripherals. It is possible to 'get into graphics' at just about any level of processing power which your budget can afford.

There are three basic approaches to 'getting into graphics': do it yourself, use a graphics software package, or obtain a laboratory information management system with integrated graphics. Each of these topics will discussed briefly.

Information Processing in Chemistry

1.5.1. Do It Yourself!

If you have near infinite amounts of time and patience, you can develop the facility for graphical display on your own. It sounds pretty simple, since there are really only two fundamental graphical instructions, MOVE and DRAW. The MOVE command picks up the pen and moves to the desired graph coordinates. The DRAW command puts the pen down and draws a straight line from the current pen location to the desired location. All components of a graph (axes, curves, symbols, labels) can be formulated as a series of MOVES and DRAWS.

The average scientist though, does not want to spend the time or effort building such a program, especially when somewhere out there, someone has already done the work and is probably offering it at an obtainable price. As in all 'do-it yourself' projects, one must weigh the actual cost of one's own time. For most of us, the best route by far is to purchase commercial graphics software.

1.5.2. Graphics Packages.

Graphical display software which has been purchased from a commercial source should allow the plotting of your data, on your devices, with your computer, with the highest speed and quality obtainable from your hardware, and the minimum user frustration. Anything less than this is inexcusable.

Graphics software is expensive, but so is your time. In addition to the time saving, you should be buying flexibility, professional quality, and plot device independence. And when it breaks, its someone else's problem to fix it.

We have had a great deal of success and satisfaction with the TELAGRAF and DISSPLA graphics packages from ISSCO (San Diego, CA). TELAGRAF is an interactive program, which employs a restricted English vocabulary to control the generation of graphs, bargraphs, piecharts, and pages of text. TELAGRAF supports nearly any graphics device in an exceedingly simply way. This means that TELAGRAF will work with whatever device is available. In addition, development work can be performed on a medium to low resolution local device and the finished product sent to a high quality, remote device.

The instructions that are used to generate an image (graph, pie, etc.) can either be entered interactively or brought in from a file. Anything that the user does not specify is supplied by default by TELAGRAF. For example, the instructions:

```
GENERATE A PLOT
TITLE IS "HYDRODYNAMIC VOLTAMMOGRAM"
X AXIS LABEL IS "VOLTAGE (VS) A(G)/C(L)"
Y AXIS LABEL IS "NORMALIZED RESPONSE"
TITLE STYLE IS COMPLEX
PAGE-BORDER IS NO
DATAFILE 'HYDRO'
***FILE***
```

were used to generate Fig. 1.

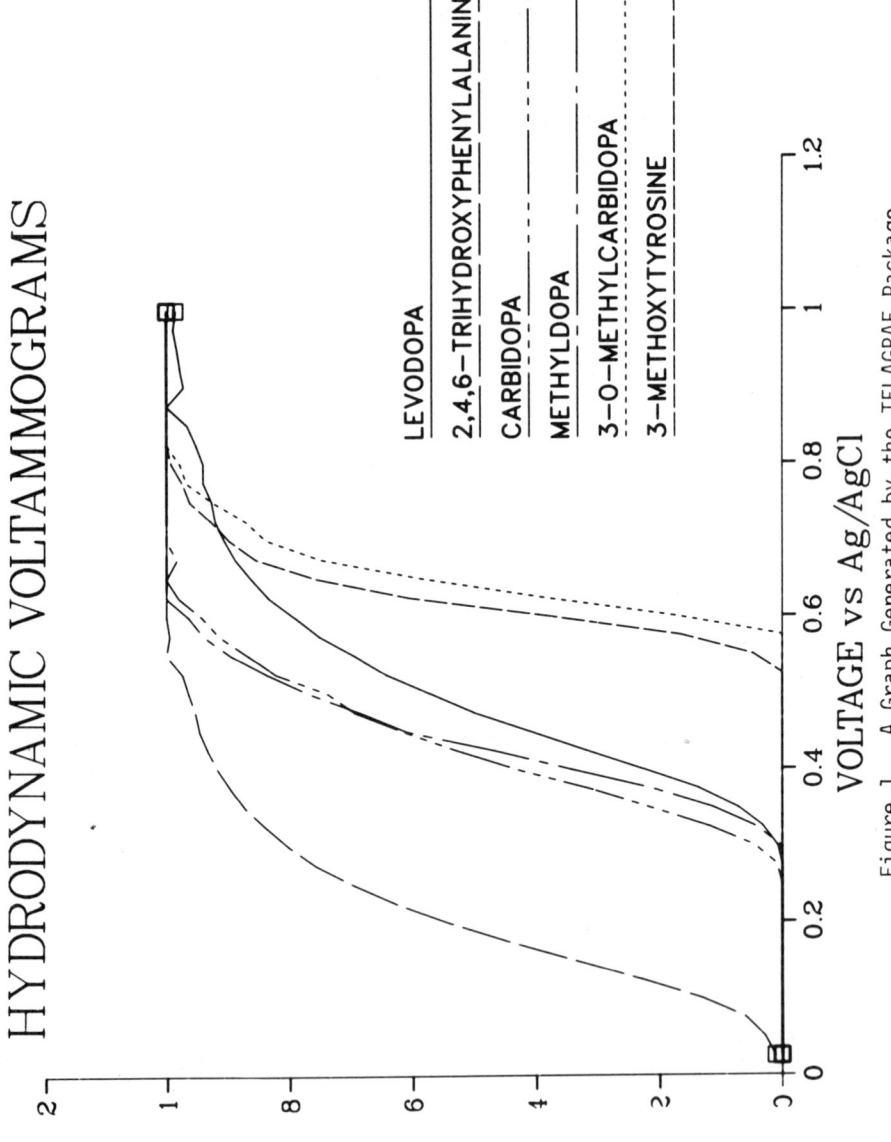

Figure 1. A Graph Generated by the TELAGRAF Package.

Information Processing in Chemistry

In some cases, a conversational program may not allow exactly what you have in mind. For this reason DISSPLA is made available. DISSPLA consists of FORTRAN-callable subroutines which perform all of the intricate tasks needed to generate graphical images. This includes axes generation, scaling, selection of character fonts and sizes, etc. These routines can be incorporated into a user's program to provide customized graphs. For example, the DISSPLA calls used to generate the plot in Fig. 2 were:

```
/INCLUDE DISSPLA        ! get DISSPLA program
 CALL TK4006 (480)      ! Tektronix terminal
 CALL BGNPL (-1)        ! begin the plot
 CALL COMPLX            ! use complex style lettering
 CALL NOBRDR            ! don't make a page border (box)
 CALL PEAK3D            ! user defined 3-D function
 CALL DONEPL (0)        ! finish the plot
```

1.5.3. Laboratory Information Management Systems.

While graphical display might be the end goal for an information processing activity, it does not represent the complete picture (pun intended). The results graphed had to come from somewhere, be it experiment or calculation. The capability to integrate the gathering, examination, and presentation of information would seem to be highly desirable. But how can this be done in a research laboratory, where the needs change on a continual basis.

The old way to use computers was to learn to program, write your own programs, and then modify your programs to suit your changing needs. As the use of computers became more extensive, and the problems attached more involved, two alternative approaches emerged. The first is have an official 'programmer' write your code. In this case you have to carefully specify exactly what you want the program to do, and how to test and evaluate its performance. Quite often the programmer will not have any understanding of the problem being addressed by the program, and will not care either.

The second approach is to buy a 'canned' program. Hopefully the commercially available software will do what you want. If not, too bad. You probably will need to buy all sorts of programs: word processing, data management, statistics, graphics, simulation, etc. Each of these programs will have its own style, commands, quirks, and inconsistencies. In addition, the programs probably will not be conveniently capable of passing information between them.

The future alternative (available today) would allow programmability when absolutely necessary, 'canned' utilities when they can be specified for permanent usage, but generally complete flexibility in a convenient, organized yet powerful manner. We have had the fortune to gain experience with such a system, namely the RS/1 package from Bolt Beranek and Newman Computer Systems (Cambridge, MA).

RS/1 (Research System One) provides an important middle ground between programming and canned application programs, which is especially geared to scientific researchers. It has been called an 'electronic laboratory notebook', and we have had a great deal of enjoyment in learning the RS/1 approach. We will take the rest of this section to briefly demonstrate the

3-D SURFACE

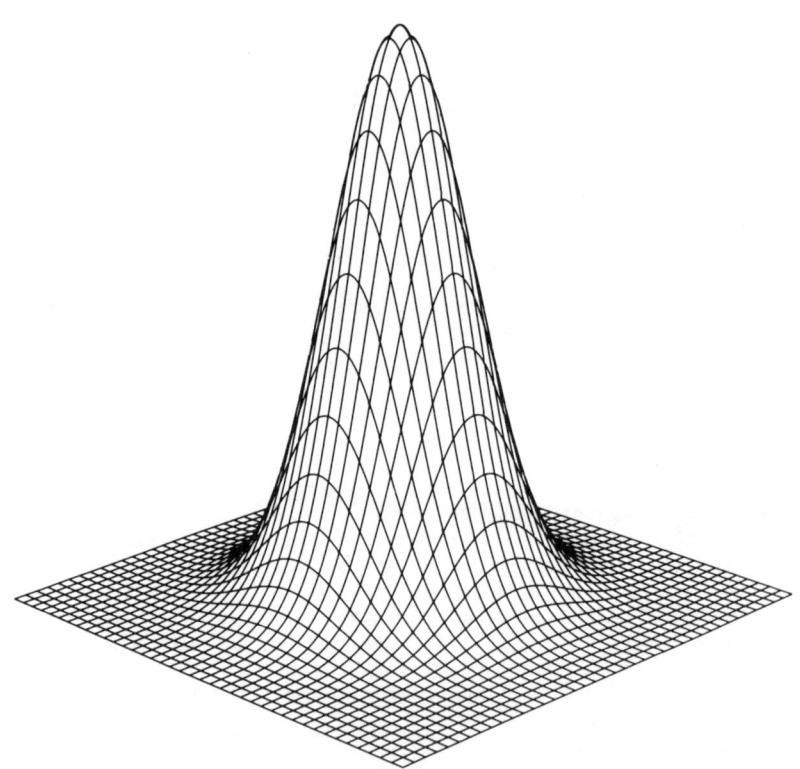

Figure 2. A Three Dimensional Surface Generated by the DISSPLA Package.

RS/1 philosophy.

The basic unifying concept of RS/1 is the table, a two dimensional place to put things, indexed by rows and columns. Data is stored in tables, the information needs to make a graph is a set of tables, the interface to peripheral devices is table driven. This prompted us to paraphrase William Shakespeare in saying that 'All the world is a table!' The other unifying feature of RS/1 is its use of a consistent command language, based on a restricted English vocabulary. RS/1 instructions generally have two modes, a 'silent' version, which minds its own business, and an interactive version, in which the user is prompted for additional information.

For example,

>   MAKE GRAPH my_graph FROM COL 2 VS COL 1 OF my_data

would generate the proper graph files, while

>   MAKE GRAPH

would initiate a user dialog:

>   NAME OF GRAPH? my_graph
>
>   TITLE OF GRAPH? An Interesting Experimental Relationship
>
>   X AXIS label? Time
>
>   X AXIS LINEAR (LIN) or LOG? [LIN]
>
>   Y AXIS label? Amplitude
>
>   Y AXIS LINEAR (LIN) or LOG? [LIN]
>
>   CURVE 1 X values: ENTER, from TABLE, of NONE (for a function)? none
>
>   F(X) = sin(10*x)/x
>
>   XLOW = 0
>
>   XHIGH = 5
>
>   YLOW = −2
>
>   YHIGH = 2
>
>   Do you want to go on and add curve 2? no

Both types of MAKE GRAPH would generate a graph like that shown in Fig. 3. Making a table, piechart, or bargraph is quite similar. The objects can also be DISPLAYed, PRINTed, PLOTted, and EDITted.

An additional capability is to WATCH a table. In this instance the table is displayed on the top portion of the terminal, and any changes that are made to the table are immediately shown. This facilitates the use of powerful table modifying commands, such as:

```
ADD ROWS TO TABLE my_table FROM ROWS 5 TO 10 OF TABLE your_table
SET ROW 5 OF my_table = ( ROW 4 + ROW 3 ) / SQRT ( ROW 2)
SET ROW 0 COL 3 OF my_table  = 'Temperature'
TRANSPOSE TABLE my_table INTO new_table
```

A table cell can contain numbers, text, computer programs or other tables. Any element of a table can be editted with the RS/1 PEN editor, or any other editor on the computer.

In addition to being able to graph an arbitrary function (to see what it looks like), as shown in the previous example, RS/1 can fit function to a set of experimental data. There are three types of curve fitting instructions:

```
FIT LINE TO CURVE 1 OF my_graph
FIT POLYNOMIAL OF DEGREE 5 TO CURVE 4 OF my_graph, and finally
FIT FUNCTION
```

which prompts the user for an arbitrary function. The result of execution of each of these commands is a complete mathematical description of the fitted function, and a statistical analysis of the goodness of fit.

In addition to the capabilities described above, RS/1 can perform statistical analysis (t-tests, analysis of variance, etc.) and modelling. A model is a table containing equations and parameters to be used in a calculation. However, in contrast to more conventional computer usage, RS/1 writes the program to perform the model's calculations.

For example Table I shows the model for a typical budget. The construct [R,C] refers to the current table cell, so [R,C-1] refers to the previous column of the current row, and [R-2,C] means two rows above the current column. The symbols '↑' and '<' mean to copy the expression in the above or left adjacent table cell into the current cell. This facilitates the use of the same formula in several places without having to type it in multiple times. Table II shows the result of the model calculation:

```
COMPUTE MODEL budget INTO budget_results.
```

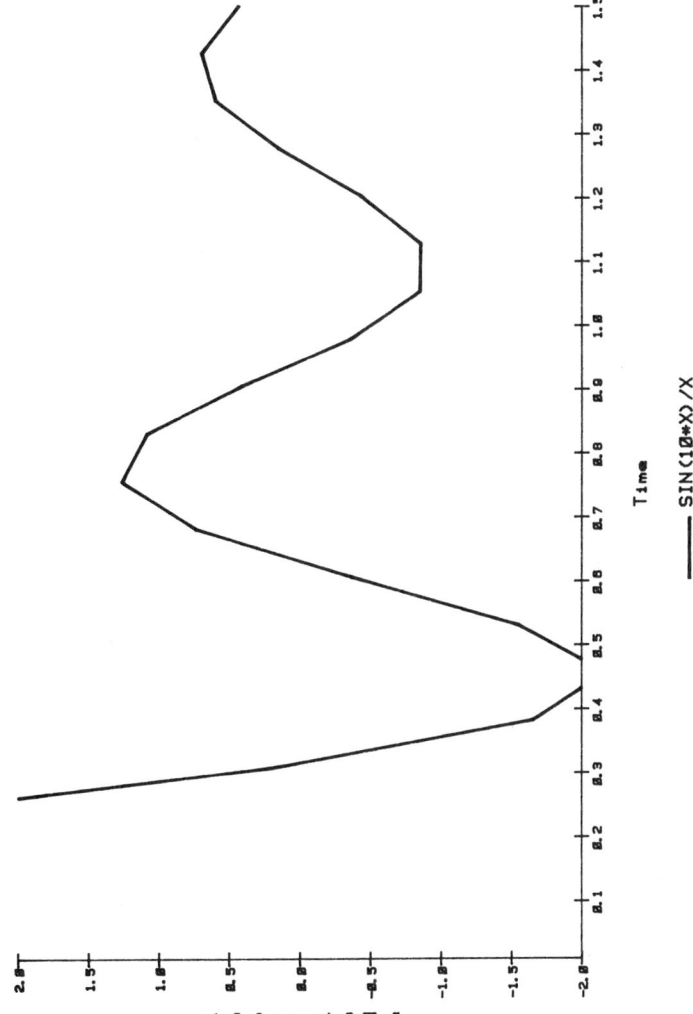

Figure 3. A Simple Graph Generated by the RS/1 Package.

The hourly rates have been multiplied by the number of hours to give expense amounts, and the sub-expenses have been summed to give the budget total. The budget can be revised by editing in new parameters, such as the overhead rate, and the model recomputed.

Finally, we would like to give a chemical example of the use of RS/1. We have a set of absorbances as a function of time for a certain chemical reaction. We would like to determine the kinetic order of this reaction. A table of the raw data was created using the MAKE TABLE dialogue. We then created two new derived columns of data for first and second order reactions:

```
SET COL 3 OF reaction_order = LOG ( COL 2 OF reaction_order )
SET COL 4 OF reaction_order = 1 / ( COL 2 OF reaction_order )
```

The resulting table is:

REACTION_ORDER 7R x 4C

Determination of the Order of a Reaction

0	1 Time (min)	2 [A] Molar	3 ln [A]	4 1/[A]
1	0	0.100	-2.302585	10.000000
2	10	0.084	-2.476938	11.904762
3	20	0.071	-2.645075	14.084507
4	36	0.054	-2.918771	18.518519
5	58	0.037	-3.296837	27.027027
6	92	0.020	-3.912023	50.000000
7	140	0.009	-4.710531	111.111111

We then made a graph for the raw data and graphs which correspond to zeroth, first, and second order reaction plots:

```
MAKE GRAPH reaction_order_g  FROM COL 1 VS COL 2 OF reaction_order
MAKE GRAPH reaction_order_g0 FROM COL 1 VS COL 2 OF reaction_order
MAKE GRAPH reaction_order_g1 FROM COL 1 VS COL 3 OF reaction_order
MAKE GRAPH reaction_order_g2 FROM COL 1 VS COL 4 OF reaction_order
```

We next fit a curve to the raw data, and straight lines to the transformed data:

```
FIT POLYNOMIAL OF DEGREE 3 TO CURVE 1 OF reaction_order_g
FIT LINE TO CURVE 1 OF reaction_order_g0
FIT LINE TO CURVE 1 OF reaction_order_g1
FIT LINE TO CURVE 1 OF reaction_order_g2
```

From the resulting graphs (Figs.4-7) it can be seen that the reaction is first order, and using the statistical report the various kinetic

Table I - RS/1 Budget Model Table

BUDGET   15R x 3C

Lab Budget for Project Delta

0	1 Rate	2 Units	3 Amount
1 Labor			
2 Sr. Scientist	21.60	1500	[R,C-2]*[R,C-1]
3 Assoc. Scientist	15.05	2000	↑
4 Technician	7.25	1000	↑
5			
6 Overhead on Labor Tot	0.58		[R,1]*(sum of col C)
7			
8 Other Direct Costs			
9 Travel	750.00	2	[R,C-2]*[R,C-1]
10 Computer	20.00	500	↑
11 Equipment	5.50	1000	↑
12			
13 Corporate Allocation	0.30		[R,1]*(sum of col C)
14			
15 Total Project Cost			SUM OF COL C

Table II
RS/1 Budget Model Results Table

BUDGET_RESULTS   15R x 3C

Lab Budget for Project Delta

0	1 Rate	2 Units	3 Amount
1 Labor			
2 Sr. Scientist	21.60	1500	32400.00
3 Assoc. Scientist	15.05	2000	30100.00
4 Technician	7.25	1000	7250.00
5			
6 Overhead on Labor Tot	0.58		40455.00
7			
8 Other Direct Costs			
9 Travel	750.00	2	1500.00
10 Computer	20.00	500	10000.00
11 Equipment	5.50	1000	5500.00
12			
13 Corporate Allocation	0.30		38161.50
14			
15 Total Project Cost			165366.50

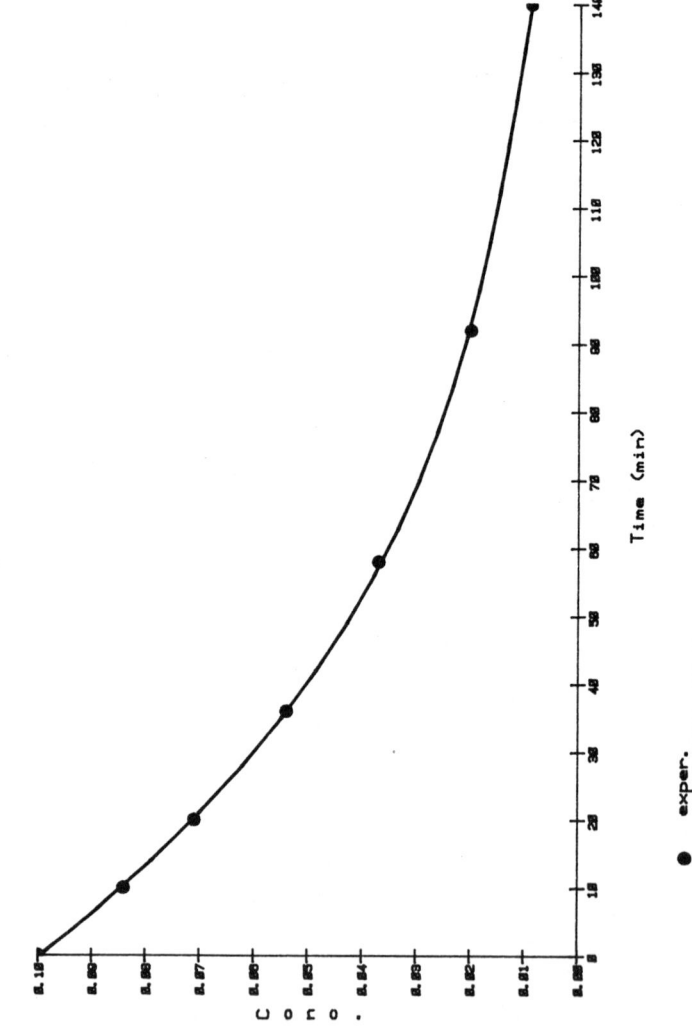

Figure 4. A Polynomial Curve Fit to Kinetic Data.

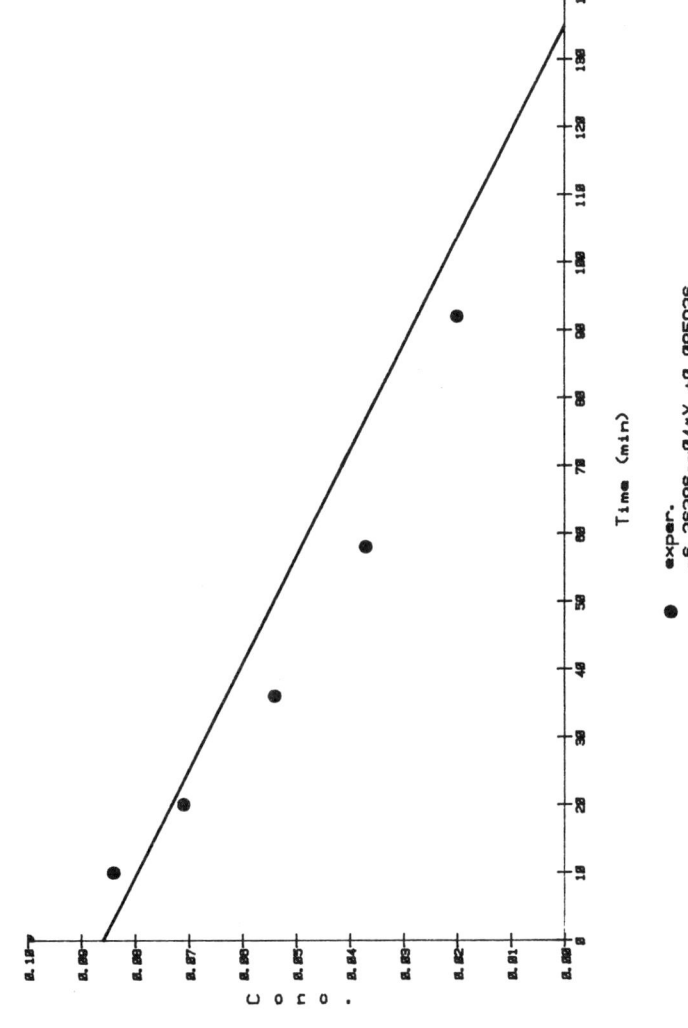

Figure 5. A Least-Squares Line Fit to the Concentration Data.

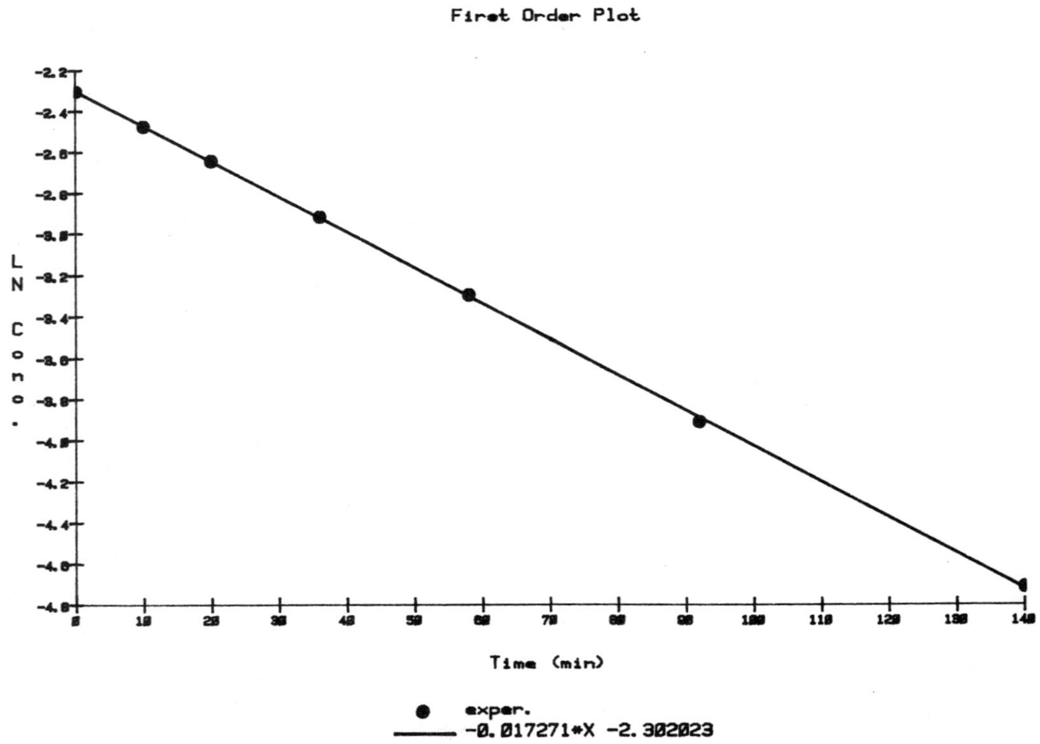

Figure 6. A Least-Squares Fit to Natural Log of Concentration Data.

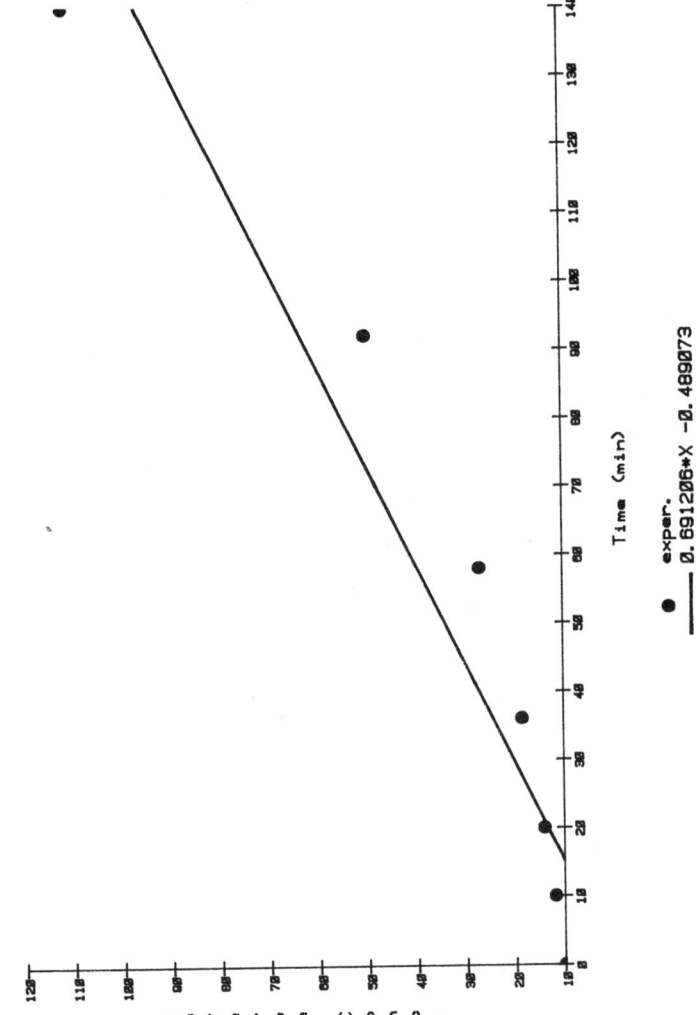

Figure 7. A Least-Squares Line Fit to Inverse Concentration Data.

parameters can be calculated.

We are just beginning to realize the power of the RS/1 approach, and feel that it will make a significant impact on the way that scientists use computers.

1.6. Bringing it All Together: Networking and Shared Resources.

Equipping a research laboratory with computers and related peripherals can be an expensive task. A computer of adequate power can be purchased for a reasonable cost but then, as in the new car market, it is the "options" that run up the price! For a computer to be of value in a laboratory, it must be able to support the various needs of the research group using it. The machine must be capable of running several software packages without bringing its processing time to a standstill. This means the scientist gets to spend research dollars on terminals and memory, that is if there is any left after buying adequate storage media. Then there are all the "neat" hardware items available such as plotters, graphics terminals, image analyzers, and line printers.

If the scientist is fortunate, he may even have funds left to purchase the software to drive all this equipment. We believe that adequate computer resources are essential in a modern laboratory environment. Sharing computer resources between research groups or departments is often the most logical, and certainly the least expensive alternative.

As stated before, the cost of a small laboratory computer is not out of the reach of most research groups. The addition of useful features such as graphics and specialized software packages are often more than one group can afford. In a university or industrial situation, funds earmarked for computer purchases can be maximized by avoiding duplication of specialized equipment. Financial cooperation between groups can result in acquisitions which would have otherwise been impossible. The goal should be to broaden the variety of equipment over a larger number of users in order to decrease costs and increase availability of important resources.

An important method of sharing resources between a large number of users is through networking. A computer network can take on many different configurations. One possibility is to maintain one central computer to which are connected several terminals. This arrangement has the advantage of concentrating both software and processing hardware in one location, keeping maintenance costs to a minimum. It is not well suited to a laboratory environment where it is necessary to interface directly to instrumentation or specialized peripherals. Another configuration is one in which several small machines are connected to a central main frame. Each individual machine would have its own storage devices and editing capabilities. Programs could be loaded up to the main frame for execution or for exchange between users. The advantage to this system is the addition of remote storage capabilities. It is similar to the previous in that it cannot address the many autonomous functions which are required in a research laboratory.

For a network to perform successfully in the laboratory, the scientist must have the means to connect into a real-time environment. A microprocessor or minicomputer should be used to control the flow of information into and out of the network. Most networks incorporate a microprocessor to control connections but a programmable interface is still necessary to accommodate file transfers and multiple device installations through a single node.

With this in mind, the last two configurations described are perhaps more suited to laboratory use. The first type incorporates a central main frame with several smaller machines connected to provide specialized functions or storage. This allows all users access to equipment which they would otherwise consider too specific for their own purchase. It also allows decentralization of facilities with the benefit to all users of being able to connect to any device on the network.

The last arrangement we will discuss in this survey is a case in which the network consists of several "independent hosts". Each node on the network is capable of controlling or accessing several devices of its own. A user on the network gains access to one of these devices by logging on to the remote node. This gives each node, or laboratory, access control over its own devices. Restrictions can be placed on certain devices or on the entire node. This arrangement is similar to the network which is in use at Boston University. Our installation consists of a broad band coaxial cable running throughout the campus. We can access directly the campus IBM 3081 mainframe as well as several minicomputers installed throughout the various academic departments. In the chemistry department we have also installed a DECnet network (Digital Equipment Corporation, Maynard, Massachusetts). Two VAX-11 super minicomputers and five PDP-11 minicomputers are currently on this intra-department network. These machines also have access to the university-wide network (Fig. 8). The advantage to this system is the control which can be exercised over exclusive departmental peripherals as well as easy access to resources outside the department.

We have discussed a variety of network configurations which we feel is representative of the installations now common. The important point of networking is the sharing of resources. We feel it is a viable solution to the problem of attaining necessary equipment with limited funding.

2. Chemically Relevant Information Processing Topics.

In the second part of this chapter we will present two specific chemical applications of information processing techniques which exemplify our philosophies, biases and interests.

There are numerous routine chemical uses of computerized information manipulation. Most of these are numerically oriented and have become, at least to us, somewhat mundane. These include curve fitting, curve smoothing, and statistical analysis. That is not to say that these are not important topics for study, and we would be the first to point out that numerical techniques have not received enough attention in mainstream chemistry. However, we are going to discuss chemical topics of a less numerical nature.

The two topics which we will cover in some detail are computerized representation of chemical structures computerized interpretation of chemical spectra.

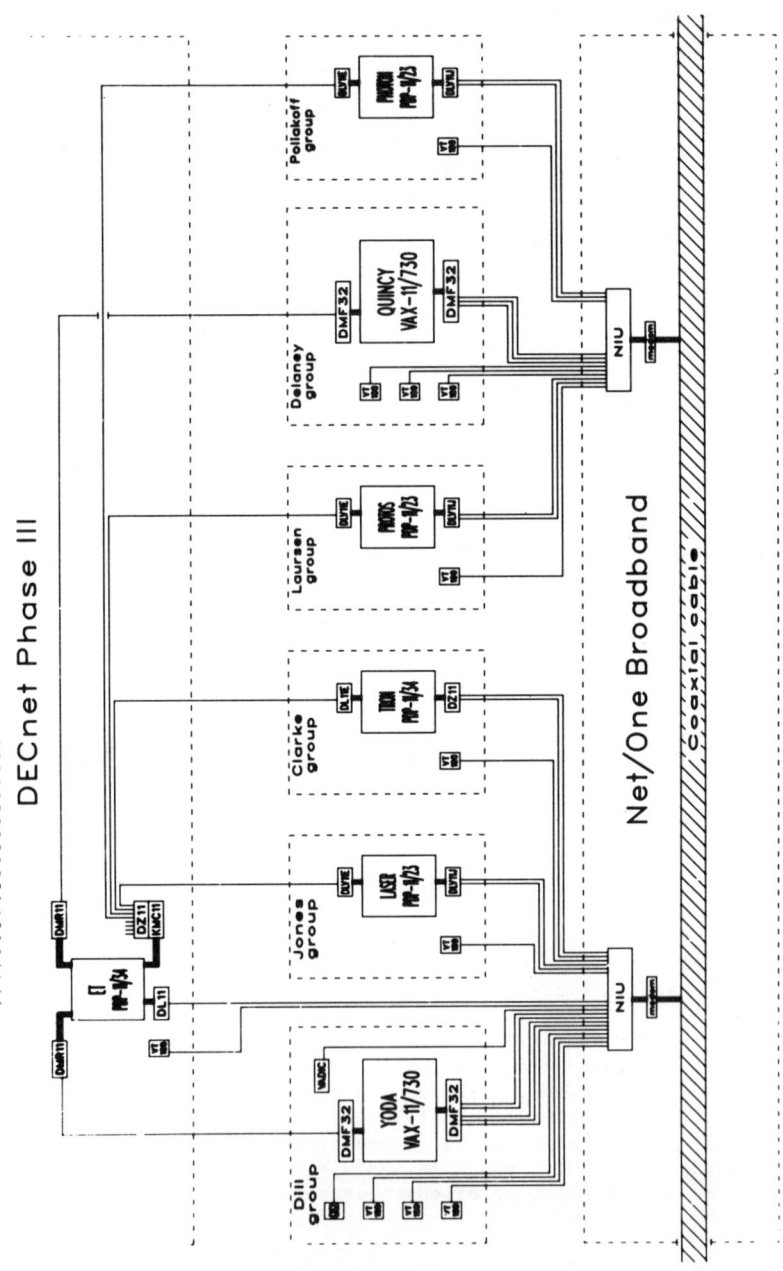

Figure 8. The Boston University Chemistry Department Computer Network.

Information Processing in Chemistry 73

2.1 Representation of Molecular Structures for Computerized Processing.

Being able to manipulate, search, and manage chemical structures is a critical capability for many chemical endeavors. The impact of computer approaches on chemical structure handling has not been as large as in other areas. Chemical compounds are most often used in a pictorial form, which is not as straight forward to accommodate as is more numerical information. However, the need for computerized structure processing outweighs the difficulties. In this section we will discuss the types of chemical endeavors which can benefit from automations in compound handling, and the tools provided by the mathematical discipline of graph theory, which makes this processing possible.

2.1.1 Uses of Chemical Structures.

We perceive four basic areas of chemistry where the ability to manipulate chemical structures by computer are vital.

o   Identification of Unknown Compounds.

   The analysis of chemical spectra with a goal of identifying unknown substances is quite common in an analytical laboratory. Modern instruments can produce spectra much faster than can be conveniently accommodated by human interpretation. The manipulation of spectral data constitutes an information processing field in itself, which will be discussed later. Interpretation of spectra 'mimicking' the human approach (artificial intelligence) has lagged behind interpretation by comparison against reference spectra (library searching) in part because the manipulation of molecular structures is not yet highly refined.

o   Organic Synthesis.

   The ability of a computer program to design an efficient synthetic scheme is under consideration by several research groups (6-9). Such computer programs could avoid some of the drudgery of planning routine synthetic routes. More importantly, a synthesis program might indicate novel approaches to making useful molecules. A computer program would be expected to consider all available synthetic information, and not be biased as a chemist might.

   Different research groups have investigated different approaches to computerized organic synthesis, but fundamental to each is the ability to conveniently represent organic compounds, and to be able to perceive a wide variety of substructures. Desirable capabilities include the characterization of rings, the perception of symmetry, and symmetrically related atoms, and the ability to express reaction transformations in a convenient language. Special languages have been developed to smooth the transfer of chemical information from the chemist to the computer.

o   'Calculation Chemistry'.

   Computerized representations of chemical compounds are frequently used by physical chemists during the computation of physical or thermochemical properties from chemical structures (10,11). The computerized formulation can also be used to reduce the complexity of the calculations by taking advantage of symmetry relationships (12).

o   Bibliographic Studies and Chemical Information Management.

Perhaps the largest application of computerized structure processing is in the usage of bibliographic data bases (13,14). Compilations of known compounds along with some of their properties can help aid in the identification of an unknown. Information such as melting and boiling points, toxicity, and reactions with characteristic reagents can help identify a given compound when little is known about its structure. The larger, and thus more complete, a data base is the better it should be expected to perform.

2.1.2 The Mechanics of Chemical Structure Handling.

A chemist usually represents the constitution of a chemical compound using a structural diagram (Fig. 9). The branch of mathematics called Graph Theory (15,16) provides a formalization of the concept of a structural diagram. In graph theory, a graph consists of a set of vertices and a set of edges, which describes how the vertices are connected to one another. To represent a chemical compound, the vertices would be atomic elements (carbon, oxygen, etc.) and the edges would be the chemical bonds.

In most of the fundamental applications of graph theory to chemistry, the hydrogens are ignored, all atoms are assumed to be carbon, and all bonds are assumed to be single. (Hydrogens are treated as though they serve about the same purpose as decimal points in numbers, they are place-holders with regard to valence/connectivity.)

It is at first rather distressing to most chemists to make these simplifying assumptions. Even though inclusion of bond order and atom type information (putting the chemistry back in) makes the theory much messier, the practice is actually greatly simplified. With the extended CM, chance similarity between two structures is much less likely. This greatly decreases the number of possibilities needed to be considered when 'matching' structures.

There are three basic problems for chemical applications of graph theory. Each problem is amenable to computerized resolution once the information has been transformed into a numerical representation. One particularly useful formalism is the connection matrix (CM), or topological matrix. The simplest type of CM is a square array, NxN, for a molecule with N atoms:

Simple Connection Table for Dichlorobenzene

0	1	2	3	4	5	6	7	8
1	0	1	0	0	0	1	0	1
2	1	0	1	0	0	0	0	0
3	0	1	0	1	0	0	0	0
4	0	0	1	0	1	0	1	0
5	0	0	0	1	0	1	0	0
6	1	0	0	0	1	0	0	0
7	0	0	0	1	0	0	0	0
8	1	0	0	0	0	0	0	0

In this CM, a '1' indicates that the two atoms are bonded together, and a '0' if they are not directly connected. The CM contains all of the

A CHEMICAL GRAPH

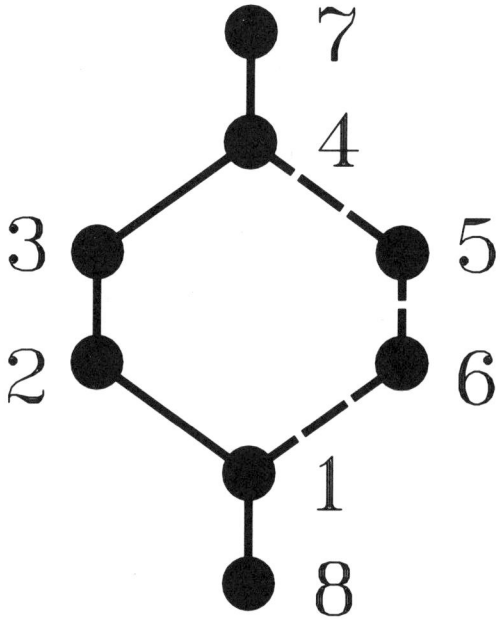

Figure 9. An Example of a Chemical Graph.

information present in the simple graph, but encoded in a numerical fashion. The type of each atom (chemical symbol, atomic number), can be included along the principle diagonal. The bond orders can be used as the off diagonal entries. For example, a '2' can be used for double bonds, and '1.5' for aromatic bonds:

Full Connection Table for Dichlorobenzene

0	1	2	3	4	5	6	7	8
1	C	1.5	0	0	0	1.5	0	1
2	1.5	C	1.5	0	0	0	0	0
3	0	1.5	C	1.5	0	0	0	0
4	0	0	1.5	C	1.5	0	1	0
5	0	0	0	1.5	C	1.5	0	0
6	1.5	0	0	0	1.5	C	0	0
7	0	0	0	1	0	0	Cl	0
8	1	0	0	0	0	0	0	Cl

Identical Structures and Canonical Numbering. The most important task is to determine if two graphs are identical (graph isomorphism). While this may sound trivial, consider multi-cyclic fused ring compounds drawn from different points of view.

Another way to formulate the graph isomorphism question is to consider the numbering, or labelling, of the vertices. If an arbitrary graph could be labelled in a clearly defined way, which depends only on the connectivity of the graph, then the isomorphism problem is solved. This is referred to as canonical numbering. Two isomorphic graphs, canonically numbered, would have identical connection matrices.

A simple, yet elegant, scheme for canonical numbering of a chemical graph has been proposed by Randic (17). (In fact, Milan Randic has been, by far, the most creative and productive chemical graph theoretician.) Randic's approach is based on considering the concatenation of the rows of a CM as a long binary number. The canonical labelling is considered to be the numbering for which this binary number is a minimum.

The mechanics of this approach, and the underlying strategy, will be conveyed by an example. Consider the simple graph in Fig. 10. If we arbitrarily label this graph, we get the following CM:

0	1	2	3	4	5	6
1	0	1	0	1	1	0
2	1	0	1	0	0	0
3	0	1	0	1	0	1
4	1	0	1	0	0	0
5	1	0	0	0	0	0
6	0	0	1	0	0	0

To get the long binary number, we place one row after the other. This produces the number:

0 1 0 1 1 0 1 0 1 0 0 0 0 1 0 1 0 1 1 0 1 0 0 0 1 0 0 0 0 0 0 1 0 0 0

An alternative notation would be to let R_i be the row of the CM corresponding to atom i. We then concatenate these row numbers to get the final results:

$R_1\ R_2\ R_3\ R_4\ R_5\ R_6$

If we label the graph differently, we get a different number. One way to achieve canonical labeling is to consider all of the possible labellings to see which produces the smallest resulting number. In general there are N! possible labellings for an N atom graph. In our case, we would need to try

$6! = 6 \cdot 5 \cdot 4 \cdot 3 \cdot 2 \cdot 1 = 720$

combinations.

A more efficient approach to make this number as small as possible, would be to start by minimizing the top rows of the graph, since these correspond to the left most part of the number. We begin by noting the degree of each vertex in the graph. The degree of the vertex is the number of edges connected to it. In our example, atoms 5 and 6 have degree, atoms 2 and 4 have degree 2 and atoms 1 and 3 have degree 3.

We then assign the first labels by giving the atom with the smallest degree the largest label number, and giving the smallest label numbers to the atoms connected directly to this atom. This insures that the top row of the CM will have as few '1's as possible, and that they are over to the right hand side. This minimizes the magnitude of the number corresponding to the top row of the CM.

This process is continued until the graph is completely labelled. However, at each stage there is a possibility of labellings which appear, at the time, equally desirable. Each of these possibilities must be considered, until they are found to be inferior to another possibility.

In our example, there are two atoms with the lowest degree (the two atoms hanging off the ring). Each of these could be assigned the label number '1'. (We can see by symmetry that it doesn't matter which we call '1', since they are equivalent.) We then assign the label number '6' to the atom bonded to atom 1. As we continue, the other atom with a degree of one is assigned the label number '2'. It is bonded to atom '5', and the last two atoms are assigned '3' and '4'. The resulting CM is:

	1	2	3	4	5	6
1	0	0	0	0	0	1
2	0	0	0	0	1	0
3	0	0	0	0	1	1
4	1	0	0	0	1	1
5	0	1	0	0	0	0
6	1	0	0	0	0	0

which yields the minimum concatenated number.

By this simple process we consider much fewer than the 6! possible labellings.

Structural Similarity and Path Codes. The second fundamental task for graph theory is the assessment of the degree of similarity between two compounds.

Chemical similarity can be defined on various levels (18). The most trivial definition would employ comparison of just the molecular formula of a compound. The next level of sophistication would be to assess the degree of functional group similarity. This approach, which we refer to as 'molecular fragmentography', has been employed by us to monitor spectral library searching performance (19). A more significant technique would be to use similarity metrics based on the conventional chemical structural diagram. Such 'connectivity metrics' use the concepts and terminology from graph theory.

Randic has developed the concept of molecular path codes (20), which have had a variety of chemical applications, including correlations between structure and chemical/physical property (21). A self-avoiding path in a molecule consists of traversing a sequence of atoms, going across the bonds, without repeating any of the atoms. An atomic path code is the number of paths of each length, starting at the desired atom. The length of a path is the number of bonds traversed. The molecular path code consists of the sum of the atomic path codes for each length and each atom in the molecule. The sums are divided by two since each bond is counted exactly twice, once in each direction. The number of atoms in the molecule is used as the first number in the molecular path code, as this corresponds to the number of paths of length zero.

An example would seem to be in order. Consider the graph in Fig. 9, corresponding to the simple hydrogen-suppressed graph of 1,4-dichlorobenzene. Table III shows the atomic path codes for each atom, and the generation of the molecular path code.

Table III - Typical Path Code Table

PATH_CODE 11R x 7C

Atom No.	length = 0	length = 1	length = 2	length = 3	length = 4	length = 5	length = 6
1		1	2	2	4	2	0
2		1	2	2	4	2	0
3		2	4	2	4	0	0
4		2	4	2	4	0	0
5		3	2	4	2	0	0
6		3	2	4	2	0	0
sum		12	16	16	20	4	0
div. by 2		6	8	8	10	2	0
path code	6	6	8	8	10	2	

Randic and Wilkins (22) used comparison of simple molecular path codes to measure similarity between a small number of pairs of related compounds. They used the summation of the square of the difference between the number of paths of each length as their metric. If the two path codes were identical the metric sums to zero, and its value increases with the dissimilarity between the path codes. Even though their approach did not employ bond order or atom type information, it was successful in producing results which were in relative agreement with a chemist's intuitive sense of similarity.

Extended 'topological indices' have been studied (23-25) which take into account the atom and bond type information in a molecule. These metrics have been employed for bibliometric compound registration and for demonstrating structure-activity relationships. We are presently studying the use of these metrics for quantitating structural similarity.

<u>Substructure Search</u>. The third chemical task addressed by graph theory is locating a molecular fragment in a molecule. This is referred to as substructure search, and is a particularly useful feature when one needs to locate all compounds in a large database which contain a certain structural fragment. Randic has shown (26) how atomic path codes can make substructure search reasonably efficient.

This subgraph isomorphism problem, as it is called in graph theoretical circles, can be stated as finding all mappings of subgraph F onto graph G. We will use the graphs in Fig. 11 as an example. For simplicity, we will use letters for the atoms in F and numbers for the atoms in G.

We begin by calculating the atomic path code for each vertex in F and G:

```
            F                           G
    atom    path code           atom    path code
    ----    ---------           ----    ---------
  A = C = D   1, 2              1 = 2     3, 2, 4
```

```
    B        3, 0         3 = 4        2, 4, 2, 2
                          5 = 6        1, 2, 2, 4
```

Note that several atoms in both F and G have identical atomic codes, due to constitutional symmetry.

We then note that for a given atom in F to be legally mapped onto a particular atom in G, the path code for the atom in G has to have every entry equal to or greater than the entry in the path code for the atom in F. This allows us to label the atoms in G that could possibly contain the atoms in F, based on the path codes. This is shown in Fig. 12. Atoms A, C, and D in F can be mapped onto every atom in G. However, atom B can only be mapped onto atoms 1 and 2 in G, because the the first entry in the path code for B is greater than the first entry in the path code for atoms 3, 4, 5, and 6.

The procedure continues by selecting an atom in G with few possible letters. For each letter on this atom, adjacent atoms are considered to see if they are labelled with letters consistent with the connectivity of F. This process greatly limits the number of possible mappings of F onto G that need be considered.

2.2 Structure Elucidation From Chemical Spectra.

In most cases, the information used to determine the structure of molecules is obtained using spectrometry, the interaction of matter with electromagnetic radiation. Although spectrometry is frequently used to obtain fundamental information about the dynamics of specific molecules, it is also routinely used as a tool for the rapid identification of the chemical structure of unknown compounds. The chemical spectrum is the profile of the intensity of electromagnetic energy absorbed or emitted, versus the wavelength or frequency of the energy. However, the interpretation of the spectrum is by no means simple or straight forward. The theoretical analysis of spectra from first principles is usually impossible, by virtue of the complexity of the calculations in all but the simplest of cases. This leaves the chemist to develop empirical relationships which relate spectral information to chemical structure, or possibly correlations which can be rationalized using the underlying theory.

An additional difficulty is the lack of complete, though complementary, information provided by the various spectrometric techniques. Some types of spectrometry excel at identifying all pieces of a molecule while others are better at discovering how the pieces are connected together. No method is unquestionably superior in all cases.

Another consideration is that the chemical spectra for a given compound are observed to vary somewhat, depending upon the experimental operating parameters. Furthermore, even if all variables are completely accounted for, some noise will always be present due to unavoidable fundamental physical limitations (Brownian motion, etc.).

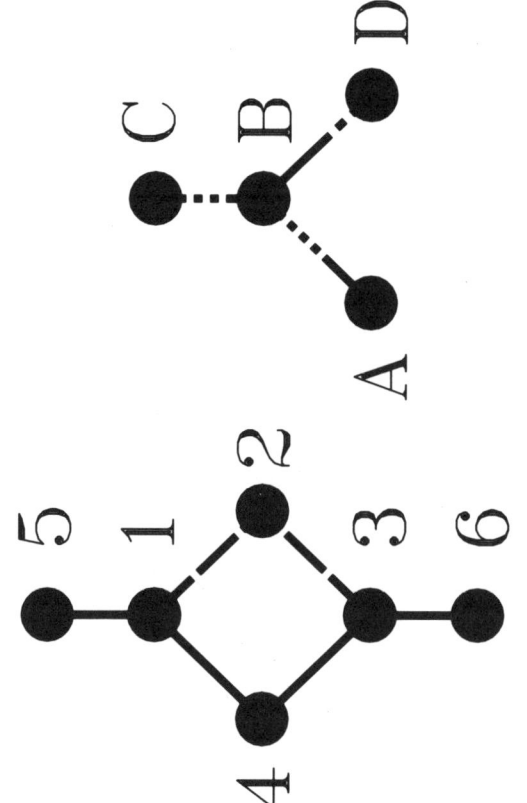

Figure 10. The Graph G.

Figure 11. The Graph F.

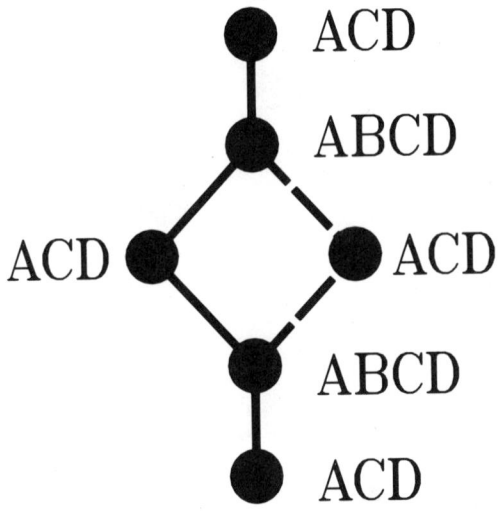

Figure 12. The Atoms of Graph Fragment G Mapped onto Graph G.

The primary goal of our laboratory is to develop a theoretical framework for optimizing and understanding the structure elucidation process. This framework will ideally facilitate a selection of the optimal methods for representing and processing the spectral information and will, futhermore, yield advances in the broader field of Information Science. In the following section, we will give an overview of the alternative approaches in computerized spectral structure elucidation, and then briefly describe our recent progress in the evaluation of library searching performance.

2.2.1 Spectral Interpretation Alternatives.

There are three major spectral structure elucidation approaches (27) each having various advantages and disadvantages and special considerations. These can be classified as deductive methods (library searching) where the range of possible chemical compounds is reduced by examining reference spectra and inductive methods (pattern recognition and artificial intelligence) in which the spectrum is interpreted to identify structural features (atoms, bonds, functional groups) present in the target molecule.

Library Searching. The most common interpretation approach is based on library searching (LS) (28) in which an unknown spectrum is compared against a reference library to find which spectra best match the unknown. Both the reference and unknown spectra are usually preprocessed to extract what is believed to be the most useful information for performing the identification. This decreases storage requirements and increases searching speed. Library searching can handle a wide range of compound types and is applicable to many instrumental approaches.

There are three main considerations in LS. First, the reference library should be large enough to encompass all the compounds that might be encountered as unknowns. This leads to very large data bases, with severe storage requirements and long search times. Secondly, the preprocessing - comparison process needs to be sufficiently immune to experimental variation that an unknown spectrum actually does match well to the proper reference spectrum. Thirdly, ancillary information regarding the unknown can be used to select a subset of the reference library in which the unknown can be expected to yield its best match. This prefiltering could dramatically increase search speed but one risks missing the right reference compound if it does not get assigned to the library subset.

Pattern Recognition. Pattern recognition (PR) has been used to classify a variety of chemical spectra (29,30), including condensed phase IR spectra (31). We have utilized hyperplane separation and the Kth nearest neighbors approach to classify vapor phase infrared (VPIR) spectra (32). In spectral analysis one generally begins with a collection of spectra of known compounds and therefore employs supervised learning PR techniques. One starts with a set of objects and a series of measurements (features) on each object. The goal is to find a way of using the measurements to separate the objects into groupings of similar constitution. This yields not only a means of assigning an unknown object to the proper class but can also indicate which measurements might actually be related to the property that defines the classes.

There are three major advantages to PR. First, the need to measure a reference spectrum of every compound which might be encountered as an unknown is eliminated, if a classifier can recognize the properties for which it was trained in spectra it has not seen before. Secondly, once the classifier is trained, the identification of unknown spectra is extremely rapid. Thirdly, the ability of PR to deal with large collections of information can give it the capability to make spectral interpretations which are difficult or impossible with simplified identification techniques.

There are two main disadvantages to PR. First, the classification of unknowns is often not perfectly correct. This prediction percentage depends on whether there is sufficient information inherent in the spectra to answer the classification question, and on how well the training set represents the unknowns actually encountered. Secondly, only binary decisions can be made, and the groupings into which compounds can be assigned are rather broad, for example functional groups. This can be partially overcome by serial or parallel cascading of classifiers but determination of single, exact structures is unlikely.

<u>Artificial Intelligence</u>. The most notable characteristic of spectral interpretation employing artificial intelligence (AI) approaches is that the process attempts to use the same strategies that a human would employ (33). This is in sharp contrast to the brute force philosophy of library searching and pattern recognition. The AI technique has several advantages. First, an AI system can utilize the existing body of interpretation developed by human specialists. This should allow an AI system to perform well from the outset, if it can utilize this information. Secondly, a functioning AI system should be able to avoid storing a library of reference spectra since the experience gained in studying known spectra ought to be embodied in the AI system. Thirdly, a functioning AI system should be able to identify unknown spectra that were not used to develop the system. This is the key property of 'intelligence'. This would eliminate the burden of acquiring reference spectra for every known compound. Furthermore, every possible compound is not, and probably never will be known.

There are two fundamental problems that an identification system which relies solely on an AI approach will need to address. First, the process that a human uses to identify spectra has not been adequately defined. To facilitate computerization of spectral interpretation, the human approach will need to be closely scrutinized, and the process detailed. Secondly, it may not be possible for the interpretation of all spectra to result in a complete and unique assignment of a molecular structure. In nearly every type of spectrometry used to elucidate molecular structure, a human specialist completes his analysis by accessing a library (in book format) of reference spectra to arrive at the final structural assignments. A human uses his experience to limit the number of reference spectra which he needs to carefully examine. However, it has been stated (34) that a skilled chemist spends fully two-thirds of his interpretation time turning the pages of a reference spectral collection.

In most applications of AI to structure elucidation to date, progress has been made by restricting the compound space to a severely small subset of known organic molecular types, generally those which are saturated, monofunctional, and non-cyclic. In such studies, large amounts of computer storage and CPU time are needed, even in simple cases. While these studies are of unquestionable value, it seems that a practically viable system is far in the future, or will be justifiable for those situations where speed is not critical, such as computerized organic synthesis (7). We propose to study the combination of inductive AI techniques and the deductive LS approach in the opposite direction from that normally envisioned.

2.2.2 Evaluation of Library Searching Performance.

Library searching (LS) remains the method of choice for computerized structure elucidation using chemical spectra (28,35). Libraries of various types of spectra have been employed, most notably mass (36), infrared (37,38) and nuclear magnetic resonance (39) spectra. In many cases, the only way to accommodate the large numbers of reference compounds needed for a useful library is to greatly compress the library spectra. The size of the library, and its effect on searching speed, can be especially critical when microcomputers are used for LS (40).

Among the most compressed spectral representations for LS, the most popular has been the use of binary intensity spectra (41). In this situation a peak in a spectrum is encoded as a '1', and the absence of a peak as a '0'. This approach has been used to advantage in mass spectrometry (MS). For vapor phase infrared (VPIR) spectrometry, we demonstrated (42) how the information content of the library and the LS performance could be improved by incorporating a measured amount of peak width information into a binary spectrum.

The comparison metric, by which similarity between an unknown spectrum and the reference library members is quantitated, is also an important criterion in designing a spectral search system for a small computer. Various metrics have been used with binary intensity representations of different types of spectrometry. The metrics are based on Boolean functions and combinations thereof.

We recently developed a procedure for the quantitative evaluation of library searching performance (42). In this approach any type of spectrometry, library representation, or comparison metric can be studied. We applied this procedure to optimize the representation of width-enhanced vapor phase infrared spectra (41).

Definition of a Spectral Representation and a Comparison Metric. A spectrum can be treated as a d-dimensional vector:

$$X_i = (x_1, x_2, \ldots, x_i, \ldots, x_d) \qquad [2]$$

where each x_i is the intensity in spectral (wavelength) channel i. In this case there are 'd' abscissa resolution elements. We can then envision each spectrum to be a point in d-dimensional space. That is, there are 'd' orthogonal coordinate axes, each one corresponding to a particular wavelength channel. In the case of a binary intensity spectral representation, each spectral element is either '0' or '1'.

A comparison metric is a function of the two spectra being compared:

$$D_{i,j} = f(X_i, X_j) \qquad [3]$$

where the result, D, is usually a measure of the dissimilarity between the two spectra. If the two spectra are identical, then D = 0, and D increases with increasing dissimilarity between the spectra. For binary intensity spectra, the simplest dissimilarity metric uses the exclusive OR function (XOR). The dissimilarity between two spectra is measured by summing, channel by channel, the result of XOR comparison:

$$D_{XOR} = \sum_{k=1}^{k=d} x_{i,k} \text{ XOR } x_{j,k} \qquad [4]$$

This is the binary analog of least squares comparison.

Generalized Evaluation of Library Searching Performance (GELS). The quantitative approach for LS performance evaluation, developed in this laboratory is described in detail elsewhere (42). In GELS, the performance of any combination of library spectra representation and comparison metric is compared against an LS standard. In our case, we employ full resolution spectra and the least-squares comparison metric as our standard.

This evaluation technique quantitatively compares the LS performance for a compressed library to the performance for the full spectra using a set of test spectra, which need not be members of the reference library. The approach measures how similar the compressed library hit-lists are to the full spectra lists.

A representative test set of compounds is selected from the standard reference library of N full-resolution spectra. Each test spectrum is searched through the N spectra library to yield hit-lists of the M best matching spectra. (N is usually large and M=10 is typical.) Each test compound is searched though the library of reduced spectra, using the comparison metric under consideration, to produce a long search list, of length N, containing each library member.

For each spectrum on the standard search lists, the list index position for the same spectrum on the corresponding long search list is found. Low list index positions indicate a high degree of similarity between the standard search lists and the compressed spectra search lists. The list index positions are summed to form a raw score. Using the best and worst possible raw scores, a normalized Figure of Merit (FOM) is calculated. The FOM allows different LS alternatives to be compared on a quantitative basis.

Hypothetical Example. A specific hypothetical example will help to clarify this evaluation process. Consider a library of 15 compounds from which a test set of 3 compounds is selected. The compounds are numbered from 1 to 15, and the test compounds are numbers 5, 10, 15. Standard search lists of the 5 best matches will be used. In this example N=15, T=3, and M=5. Table IVa shows the standard search lists for the full resolution spectral library. As expected, since each test compound was drawn directly from the library, it is found at the very top of the standard search list. Table IVb shows the long hit-lists for the library cf reduced spectra. In both

Information Processing in Chemistry

Tables IVa and IVb each column corresponds to one of the test compounds. For Table IVb there are 15 entries in each column, one for each library member. The best matching spectrum is in row 1, the second best matching spectrum is in row 2, etc.

Table IVc shows the corresponding list index positions. These are found by locating each compound of the standard search list in the corresponding long search lists (Table IVb) and noting the list position. For example, on the standard search list (Table IVa) test compound one's fourth best match is compound number 8. On the long search list (Table IVb) for test compound one, compound number 8 is found to be in list position 14. Therefore, in the list index position array (Table IVc) for test compound one, there is a 14 in column one, row 4. The rest of this list index position array was completed in an analogous fashion.

The list index position array is now used to generate a figure of merit (FOM) which can be used to quantitatively compare the results of this (hypothetical) LS system to the results of other systems. First, column sums are obtained for each of the columns in Table IVc. These sums are then added to give S, the raw index position score. The FOM is calculated, using the best and worst possible sums. The worst possible score for this example would occur when the sequence 1, 15, 14, 13, 12 is in each of columns 1-3. The first entry in each column must be a 1 since each test set member is actually in the reference library. The other four entries indicate that the compounds found high on the standard search list are simultaneously found at the very bottom of the long search lists, which corresponds to the worst possible library searching performance. According to this procedure values of S=73, B=45, W=165 and a FOM of 76.7% are obtained.

<u>Actual Example: The Effect of Spectral Intensity Resolution</u>. We have recently been studying the effect that differing amounts of intensity information has on library searching performance. Using a library of vapor phase infrared spectra (41), we created alternative libraries which ave differing amounts of spectral intensity resolution.

The amount of intensity information is measured in bits (binary digits). One bit of intensity means that there are only two possible intensity values, '0' and '1'. With ten bits of information, 1024 intensity levels are possible, since 2^{10} = 1024.

The use of less intensity information would allow more reference spectra to be stored in the same amount of space. Also, the library searching speed would be greater for a smaller library.

Using the GELS approach, we observed that nearly all of the test spectra achieved optimal performance when only 3 bits of intensity were used (Fig. 13). Additional amounts of intensity were not needed.

Table IV
An example of the summed list position
library searching evaluation.

a. STANDARD SEARCH LISTS

Search List Position	Test compound number		
	1	2	3
1	5	10	15
2	2	7	13
3	12	1	12
4	8	9	3
5	9	5	1

b. LONG SEARCH LISTS

Search List Position	Test compound number		
	1	2	3
1	5	10	15
2	2	7	6
3	12	1	12
4	4	9	13
5	3	5	1
6	7	4	10
7	10	11	8
8	15	13	14
9	13	15	2
10	6	6	7
11	9	8	5
12	14	2	4
13	11	12	9
14	8	14	3
15	1	3	11

c. LIST INDEX POSITION ARRAY

Index Position	Test compound number		
	1	2	3
1	1	1	1
2	2	2	4
3	3	3	3
4	14	4	14
5	11	5	5
Column Sum:	31	15	27
Raw Score:	73		
Figure of Merit:	76.7%		

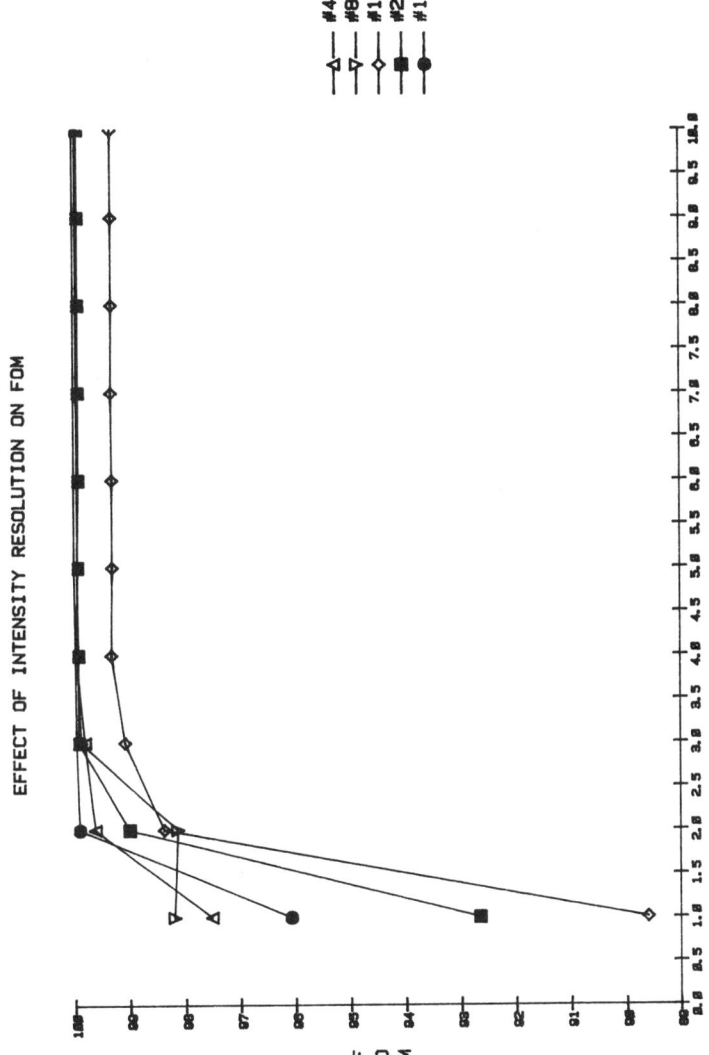

Figure 13. The Effect of Intensity Resolution on Figure of Merit.

3. Conclusion.

We have given several examples of the types of facilities available for incorporating information processing into scientific research. The number of tools available is constantly increasing along with the many uses to which these tools can be applied. We have tried to illustrate the utility of information processing and the importance of involving the scientific community with its future growth and development. The concepts and applications embodied in this field have value for all of us. We should strive to take full advantage of its uses and effect our own enhancements and refinements upon it.

4. Acknowledgement.

We would like to thank Dan Dill and Peter Dittman of the BU Chemistry Department for providing the Network Diagram (Fig. 8), and for helpful discussions. Our thanks also go to Diane Gsell for straightening out some of our English.

Acknowledgement is gratefully made to the National Science Foundation's Information Science and Chemistry Divisions (Grant No. IST-8120255) for the partial financial support of this research.

5. Literature Cited.

1. B. W. Kernighan, D. M. Ritchie, 'C Programming Language', Prentice-Hall, Englewood Cliffs, NJ, 1978.

2. D. P. Friedman, 'The Little Lisper', Science Res. Assoc., Chicago, IL, 1974.

3. R. E. Dessy, ANAL. CHEM., 55, 651A and 756A (1983).

4. H. B. Woodruff, G. M. Smith, ANAL. CHEM., 52, 2321 (1980).

5. E. J. Corey, A. K. Long, J. Mulzer, H. W. Orf, A. P. Johnson, and A. P. W. Hewett, J. CHEM. INF. and COMP. SCI., 20, 221 (1980).

6. A. K. Long, S. D. Rubenstein, L. J. Joncas, 'A Computer Program for Organic Synthesis', in Chem. and Eng. News, 19(61), 22 (1983).

7. see 'Computer Assisted Organic Synthesis', W. T. Wipke, ed., Amer. Chem. Soc., Wash. DC, 1978, and refs. therein.

8. W. T. Wipke and T. M. Dyott, J. AMER. CHEM. SOC., 96, 4835 (1974).

9. T. H. Varkony, R. E. Carhart, and D. H. Smith, in 'Computer Assisted Organic Synthesis', W. T. Wipke, ed., Amer. Chem. Soc., Wash. DC, 1978.

10. G. Klopman, M. McGonigal, M., J. CHEM. INF. and COMP. SCI., 21, 48 (1981).

11. G. W. Adamson, D. Bawden, J. CHEM. INF. and COMP. SCI., 20, 242 (1980).
12. M. Randic, G. M. Brissey, C. L. Wilkins, J. CHEM. INF. and COMP. SCI., 21 52 (1981).
13. R. Fugmann, J. CHEM. INF. and COMP. SCI., 22 118 (1982).
14. J. M. Barnard, M. F. Lunch, S. M. Welford, J. CHEM. INF. and COMP. SCI., 22 160 (1982).
15. F. Harary, 'Graph Theory', Addison-Wesley, Reading, MA, 1969.
16. N. Deo, 'Graph Theory with Appl. to Eng. and Comp. Sci.', Prentice-Hall, Englewood Cliffs, NJ, 1974.
17. M. Randic, CHEM. PHYS. LETT., 42, 283 (1976).
18. T. H. Varkony, Y. Shiloach and D. H. Smith, J. CHEM. INF. and COMP. SCI., 19, 104 (1979).
19. J. Kwiatkowski, and W. Riepe, FRES. Z. ANAL. CHEM., 302, 300 (1980).
20. M. Randic, G. M. Brissey, R. B. Spencer, and C. L. Wilkins, COMPUT. CHEM., 3, 5 (1979).
21. M. Randic, J. AMER. CHEM. SOC., 97, 6609 (1975).
22. M. Randic and C. L. Wilkins, J. CHEM. INF. and COMP. SCI., 19, 31 (1979).
23. L. B. Kier, J. PHARM. SCI., 69, 1034 (1980).
24. L. A. Evans, M. F. Lynch and P. Willett, J. CHEM. INF. and COMP. SCI., 18, 146 (1978).
25. T. DiPaolo, L. B. Kier, and L. H. Hall, J. Pharm. Sci., 68, 39 (1979).
26. M. Randic and C. L. Wilkins, J. CHEM. INF. and COMP. SCI., 19, 23 (1979).
27. M. F. Delaney and P. C. Uden, J. CHROM. SCI., 17, 428 (1979).
28. H. S. Hertz, R. A. Hites, and K. Biemann, ANAL. CHEM., 43, 681 (1971).
29. P. C. Jurs and T. L. Isenhour, 'Chemical Applications of Pattern Recognition', Wiley, NY, 1975.
30. P. C. Jurs, B. R. Kowalski, T. L. Isenhour, and C. N. Reilley, ANAL. CHEM., 42, 1387 (1970).
31. B. R. Kowalski, P. C. Jurs, T. L. Isenhour, and C. N. Reilley, ANAL. CHEM., 41, 1945 (1969).
32. M. F. Delaney, P. C. Denzer, P. C. Uden, and R. M. Barnes, ANAL. CHEM., 12A, 963 (1979).

33. M. Penca, J. Zupan, and D. Hadzi, ANAL. CHIM. ACTA, **95**, 3 (1977).

34. T. Hirschfeld, Federation of Anal. Chem. and Spectrosc. Soc. meeting, Phil. PA, Oct. 1980.

35. G. T. Rasmussen and T. L. Isenhour, J. CHEM. INF. and COMP. SCI., **19**, 179 (1979).

36. G. T. Rasmussen, T. L. Isenhour, and J. C. Marshall, J. CHEM. INF. and COMP. SCI., **19**, 98 (1979).

37. J. A. de Haseth and L. V. Azarraga, ANAL. CHEM., **53**, 2292 (1981).

38. M. F. Delaney and P. C. Uden, ANAL. CHEM., **51**, 1242 (1978).

39. H. B. Woodruff, C. R. Snelling, C. A. Shelley, and M. E. Munk, ANAL. CHEM., **49**, 2075 (1977).

40. A. P. Uthman, J. P. Koontz, J. Hinderliter-Smith, W. S. Woodward, C. N. Reilley, ANAL. CHEM., **54**, 1772 (1982).

41. S. L. Grotch, ANAL. CHEM., **42**, 1214 (1970).

42. F. V. Warren and M. F. Delaney, APPL. SPECT., **37**, 172 (1983).

43. M. F. Delaney, F. V. Warren, and J. R. Hallowell, ANAL. CHEM., **55**, (Sept. 1983).

CARL TRINDLE, MICHAEL BROWN, AND M. GARY NEWTON

CHAPTER 4

Use of Algebraic Symbol-Manipulation Programs in Chemical Research and Education

I. Introduction

The computer is the latest in a long series of mechanical aids to thought. Consulting either world history or our own experience, we can list some of its predecessors:

Fingers and toes
Abacus
Slide rules
Pocket calculator

In all these mechanical aids, objects represent numbers. Manipulation of the objects takes the place of abstract numerical operations. The objects may be literal digits (fingers!), beads, or special marks. The objects may be "figurative digits" or patterns of light or magnetization. In any event, the mathematical operations are reduced to physical operations, the results of which are interpreted (decoded) in ways spelled out in unambiguous rules. The objects are physical realizations of symbols, and the operations are well defined manipulations of symbols. There is no reason why symbol manipulation must be confined to arithmetic operation! Numerology is a small (however useful) part of math; but algebraic simplification, differentiation, even integration, are all examples of perfectly analogous manipulations on symbols.

The stress on the computer as a very effective calculator has obscured its versatility, and general applicability as a symbol manipulator. However, algebra and calculus programs have existed for almost as long as programmable machinery has existed [1]. Symbol manipulation by a program was pointed out by the very first expert in software design, Augusta Ada Byron, Countess of Lovelace [2]. She said that the computer could

> arrange and combine its numerical quantities exactly as if they were letters or any other general symbols; and in fact might bring out its results in algebraic notation, were provisions made accordingly.

In contrast to strictly numerical computation, which can be done sometimes better by idiot savant or by bones than by the most brilliant of mathematicians, algebra, geometry, calculus and higher mathematics seem to require the spark of human intelligence. In the heady fifties, when computers were first turned away from preparing mathematical tables, describing ballistics, and decoding military and diplomatic messages toward the task of aping human intelligence, investigators asked, "What is unambiguously intelligent behavior?" Chess mastery; translation of natural languages (especially Russian); and higher mathematics were three nominations for such "essences of intellect" [3].

Among the most dramatic early triumphs of the effort to capture in electronics whatever abilities were needed to do calculus was a program called "SAINT" [4]. This program earned an 85% - B - on the MIT first year calculus exam. (After it was taught the trick of partial fractions, it scored 100%). Of course like many another disabled "student" it required some special aid in communication.

Since then the ability and speed of symbol-manipulating programs have increased considerably, aided both from dramatically improved hardware and (to a lesser degree) better algorithms. Programs to do calculus are available for the CDC (REDUCE [6]) and for the IBM (FORMAC [7]). There is even a well designed microcomputer package (MuMATH [8])! There are many special-purpose programs as well, for quantum electrodynamics, relativity, high-energy physics, and celestial mechanics [9]. Chemistry does not seem very well represented, perhaps because theoretical chemists have been traditionally oriented toward numerical methods. Even group theory, a prime candidate for treatment by symbolic manipulation programs, has been dealt with numerically. (Examples include IRREP, GPTHEORY, and PROJECT, all distributed by QCPE [10].) Why not use these programs more, whenever tedious algebra is a stumbling block?

II. Some Pros and Cons of Computer Algebra

To the question, "why have these programs been so neglected in chemistry?" there are several possible answers:

1. These programs are expensive; if not in my money, then in my time.
This is true. But this will change. Of course any new technique requires some time to master. Only if there is a long range savings of time should you invest your time now. Time will be saved if you ever encounter these situations:
You are using a table of integrals. You have reason to suspect a "typo." Several tables give differing results.
You are reading the theoretical section of a paper. The author remarks, "After considerable manipulation, we find . . . "
You have developed a theoretical result, repeatedly. The result Monday has a factor of 8; Friday's result doesn't.

2. Doing the algebra is a way to understand the details of a problem. By assigning this work to a machine I lose intimate contact with the math. (Or, variously) computer algebra would be fine for big problems; but I choose small, feasible problems for pedagogic value. Here the details help to convince the student that the derivation (or calculation) is honest and believable.

These are two kinds of objections to mechanization of algebra, which are really paraphrases of one another. One certainly achieves a thorough understanding of a calculation by seeing every detail and the progression in slow degrees from the presumptions of a model to testable numerical results. However, there is always a danger that the richness of detail is distracting, and that the point of the journey is lost. In practice, we do not shrink from delegating the strictly arithmetic work to a calculator or computer; what makes algebra or calculus different? Somehow we believe that the exercise is uniquely convincing, that the laborer will "see for himself" rather than through the eyes of others. There is a stage where we "leave the proof" to the reader. This delegation does not promise that the reader will gain any new insight not already common, but that he will have made the derivation somehow his very own. This argument has echoes. It sounds much like the great - and now forgotten - pocket calculator debate. No one now believes that manual arithmetic yields much insight into any kind of calculation. In fact it is a barrier to understanding, being entirely mechanical, error-prone, and in a sense, beside the point. It is entirely possible to make math repulsive by dwelling on such detail.

Is algebra - is calculus - comparable in these respects to arithmetic? Obviously my opinion is that they are. But one must distinguish the modeling procedure which produces the equations and integrals which are to be simplified, solved, or evaluated, from the procedures by which such re-expressions are accomplished. Mechanizing algebra and geometry will leave more time for the creative modeling step, and permit more thorough analysis of the results of the model. Mechanical aids are useful only for the latter (at present).

Use of Algebraic Symbol-Manipulation Programs

III. Getting Started with Computer Algebra

Let us consider more thoroughly the very real problem of getting started. Even those who are comfortable using the computer as a numerical aid, and are skilled FORTRAN programmers, will find the transition to computer algebra more than trivial. The major obstacle is the expression of these programs in the de facto Esperanto of artificial intelligence research, the list processing language LISP (or closely similar dialects) [11]. LISP is a language of relationships rather than values, and requires a considerable change in the FORTRAN /Pascal /BASIC programmer's point of view. To ease the transition, most computer algebras have a calculator mode. (Most of the examples given below are generated in this mode, with the aid of the microcomuter package MuMATH.) After some preliminary grappling with the operating system, one is prompted by some special symbol ("?" In the examples). One responds by defining expressions referring to one or more "atoms", which are the fundamental units subject to manipulation. For example, we define the generating function for the Hermite polynomials:

? Gm: #e^(x^2-(m-x)^2);   [$\exp(x^2-(m-x)^2)$]

Here we have defined the expression Gm as a function referring to atoms x and m. The notation #e refers to the base of natural logs. Similar notation (#i, #pi) identifies common constants. The generating function for the Hermite polynomials contains all information necessary to the definition of these polynomials, which play a significant role in the oscillator eigenfunctions in quantum chemistry. The k-th Hermite polynomial can be recovered from gm by differentiating k times and evaluating the limit of the result as m tends to zero. This is accomplished in MuMATH as follows:

h0:gm;	Define the zeroth polynomial
evsub(h0,m,0);	Set m to zero
h1:dif(h0,m);	Differentiate with respect to m
evsub(h1,m,0);	Set m to zero, defining the 1-st order form
h2:dif(h1,m);	Repeat as necessary (etc.)

You may find it preferable to use a recursion relation:

    h0:1;   The zero-th polynomial
                             is defined
    h1:2*x;   as is the first
    h2:2*x*h1-2*1*h0;   and the second.
    h3:2*x*h2-2*2*h1;   The recursion relation
                             defines the third.

A general term might look like this:

    hnpl:2*x*hn-2*hnml

Already it becomes clear that programming is a desirable thing to learn, particularly given the full power of recursion in LISP or MuMATH.

```
        FUNCTION HERMITE (X,N)
            WHEN N=0, 1 EXIT,
            WHEN N=1, PROD (2,X) EXIT,
            DIFFERENCE (
            PROD(PROD(2,X),HERMITE
                    (X,DIFFERENCE(N,1))),
     PROD (PROD (2,DIFFERENCE(N,1)),
            HERMITE (X,DIFFERENCE(M,2))))
            ENDFUNS
```

As a more advanced example, consider the use of these generating functions to evaluate harmonic oscillator overlap integrals: we use the generating function, g(n,x).

    ? Gn:#e^(x^2-(n-x)^2);

    @ #e^(x^2-(-x+n)^2)

The "@" signals a computer response.

    ? Gm:#e^(x^2-(-x+m)^2;

    @ #e^(x^2-(-x+m)^2)

    ? Grand:gm*gn*#e^(-x^2);

Use of Algebraic Symbol-Manipulation Programs

```
@  #e^(x^2-(-x+n)^2-(-x+m)^2)

?  Grin:int(grand,x);

@  #e^(2*m*n)*3pi^(1/2)*erf(z)/2
```

The system has integrated the Gaussian form. Here z is the expression

x-n-m.

And the definite integral over infinite limits becomes:

```
#e^(2*m*n)*#pi^(1/2)/2.
```

We may recover individual overlap integrals from this expression by consulting the equivalent form:

```
?  Grin2:sum(n,m)(int(hn*hm*#e^(-x^2),x))
              /(fact(n)*fact(m)
```

from which we recover the specific integrals S(K,L) by the operations:

1. K differentiations with respect to m
2. L differentiations with respect to n
3. Limit as m approaches zero
4. Limit as n approaches zero

All of these operations are straightforward with MuMATH. However, we leave those details in order to consider a more challenging problem: calculating Franck-Condon factors for states with different vibrational frequencies and equilibrium distances. The general approach is very similar. We define

```
?  Gm:#e^(x^2-(m-x)^2);
?  S:px+q
```

The variable s incorporates the different position (q) and the different frequency (p). Then

```
?  Gns:#e^(s^2-(n-s)^2;
```

is the "shifted" generating function. The integrand is:

? Grands:#e^(-x^2/2)*#e^(-s^2/2)*gm*gns;

The computer responds

@   #e^(4*n*x*p+4*n*q+4*x*m-2*x*p*q -x^2*p^2-2*n^2-x^2-2*m^2-q^2)/2

We complete the indefinite integration:

? Grins:int(grands,x);

@ (Output is complicated)

We attempt to simplify the expression by successive expansion and factorization.)

? Grins:expd(grins);

@ (output is complicated)

? Grins:fctr(grins);

@ #e^((8*n*m*p+4*n*q-4*m*p*q+2*n^2*p^2
    -2*m^2*n^2*p^2-2*n^2 -2*m^2 -q^2)/
    (2 +2*p^2))*#pi^(1/2)*erf(z)/(2+2*p^2)

The expression z is essentially a linear form z = a*x+b. The limit of erf(z) is 1. The coefficient of erf is the sought expression, from which we recover specific integrals by the successive differentiations and limits, as above. Analysis of the integral shows that maximum overlap occurs when:

1. N=m and q=0, or
2. Q=n-m

As a bonus of the exercise, FORTRAN or BASIC code for the integrals is easily generated by editing the MuMATH output.

IV. Applications in Research

The illustration in part II could have arisen in a first quantum chemistry course (with the Franck-Condon integral assigned as a problem). Even the rather limited ability of MuMATH can be valuable in research, however. In this section we view a few specific examples where MuMATH has helped ease some problems in my own research. These examples will be somewhat abbreviated to conserve space.

A. A Measure of Basis Set Quality

The closure property of a basis can be expressed in the form:

fsum:sum(k)[(f(k:x)*f(k,a)=delta(x-a)]

where k is an index and x, a are values of the coordinate. Any practically useful basis set will be finite, and the delta function will be represented approximately. We can measure the quality of the basis, roughly, by the sharpness of the basis set's representation of the delta function. A measure of such sharpness can be the parameter b in the function of such sharpness can be the parameter b in the functional form

gauss:(p*pi/b)^(1/4)*#e^(-b*(x-a)^2);

if b is chosen to match the basis set representation in the integral-least-squares sense:

grand:(fsum-gauss)^2
match:int(grand,x);

We wish to chose b so that match is a maximum:

maxmatch:dif(match,b);

This occurs when maxmatch vanishes. Numerical solution of the equation maxmatch=0 will be necessary, but the integrals

gpro:int(f(k,x)*#e^(-b**x-a)),x)

can be done for any common basis, where f(k,x) may be exponential or Gaussian, by MuMATH.

B. Lineshapes for P-31 and H-2 NMR Spectra in Lipids

It is possible to represent NMR lineshapes in lipids by a Lorentzian form [12] where the damping factor depends on the angle between the C-D bond vector and the normal to the lipid surface: call the angle theta.

```
P2:(3*x^2-1)/2
```

P2 is the second Legendre polynomial; x=cos(theta)

```
gamma:a+b*p2
```

```
shape:f/(f^2+gamma^2)
```

The defined shape would obtain only for parallel C-D vectors. We must average over angles to find the observed shape:

```
obshape:int(shape,x)
```

Although fitting the theoretical form to observed lines becomes a numerical non-linear least-squares problem, MuMATH saves computer time by supplying an analytical form for obshape. The analytic form had been overlooked until 1980. [13] Numerical quadrature is still practiced in many groups.

C. Molecular Geometry Calculations

It is a straightforward (numerical) exercise to compute Cartesian coordinates, given either X-ray results or given the set of primary bond distances, valence angles, and torsion angles [14]. However the usual algorithms sometimes fail to preserve symmetries and to close rings perfectly. In addition, they do not ease the visualization of local ring modes, namely puckering and pseudorotation.

Cremer and Pople [15] have devised puckering coordinates specific to ring motions, which apply to irregular rings as well as regular polygons. For the ring in question, the first step is to define the mean plane. Let the position of each atom be specified by a radius vector $\underline{R}$: set the origin at the centroid:

sum(j) [$\underline{R}$ (j)]=0.

Define vectors $\underline{R}'$ and $\underline{R}''$

$\underline{R}'$ = sum(j) [$\underline{R}$(j)*sin(2*pi*(j-1)/N)]
$\underline{R}''$ = sum(j) [$\underline{R}$(j)*cos(2*pi*(j-1)/N].

Then the unit vector

$\underline{n}$ = $\underline{R}'$ x $\underline{R}''$ is normal to the mean plane.

Departures of each atom from that mean plane are given by

z(j) = $\underline{R}$(j) (dot) $\underline{n}$.

These z values can be reproduced by a set of amplitudes q and angles phi.

z(j)= sum(m) [(2/N)]^(1/2)*q(m)*
       cos(phi(m)+2*#pi*m*(j-1)/N)

Pseudorotation and puckering motions are naturally described in these coordinates, and conformations such as boat, twist-boat and chair occur at simple values of the angles. While these names suffer some ambiguity when the rings are composed of irregular line segments, the coordinates are general. It becomes convenient to case experimental (crystallographic) coordinates for rings into Cremer-Pople coordinates, and to express the potential energy of ring puckering in this new form as well. Given such a potential, it might be of interest to recover wave functions describing puckering. Our very first step must be to transform the Schroedinger kinetic energy operator into Cremer form. Here computer algebra can be of some assistance; although MuMATH was not capable of solving some aspects of the transform, it was of some help in leading me away from some false starts. Some preliminary analysis was required.

From the definition of q, phi and the chain rule, we find:

    dif(z) = sum(m) [der(qm,z)*dif(qm)]
    + sum(m)[der(phim,z)*dif(phim)]

The notation: "der (y,x)" is to be read "the derivative of y with respect to x"; "dif(x)" is to be read "the differential (operator) with respect to x".

One must solve some simultaneous equations to obtain the derivatives. The equations result from partial differentiation of the equations:

    q(m)*cos(phi(m))=sum(j) [z(j)*cos(2*#pi*m*(j-1)/N)]
    q(m)*sin(phi(m))=sum(j) [z(j)*sin(2*#pi*m*(j-1)/N)]

We find that

dif(z)=sum(m) [cos(phi(m) + 2*#pi*m*(j-1)/N)*dif(q(m))
    + q(m)*sin(phi(m)+2*#pi*m*(j-1)/N)*dif(phi(m))]

MuMATH provides help in the algebraic solution of systems of equations (through its symbolic matrix inversion) but the capability is limited to problems (such as this one) which are simpler to solve directly by inspection. Nonetheless MuMATH does the job, and provides training in the system and proof against carelessness.

The next problem is to construct the Laplacian operator, del-sq. In the event, we are obliged to evaluate sums of the form:

    sum(j) [TR(phi(m)+2*#pi*m*(j-1)*
    TR(phi(l)+2*#pi*l*(j-1)/N)]

Here TR can be either sin or cos. These sums are known [16], but can be verified by MuMATH. The program is capable of expanding trig functions, replacing products by the proper sums, and will replace trig functions by the complex - exponential equivalents. With the aid of these transforms, simple finite geometric series are obtained. MuMATH can sum these directly, though it is not capable of extracting the geometric series from involved expressions. That becomes the task of the operator.

After sums are performed, the Laplacian operator becomes:

sum(m) [der((dif(q(m))q(m) + (1/q(m))*dif(q(m))
     + (1/q(m)^2)*der((dif(phi(m)),phi(m))

If there is no restraining potential, the Schroedinger equation is solved by Bessel functions in q(m) and complex exponentials (angular momentum functions) in phi(m). If there is an harmonic potential k(m)*q(m)^2, then the eigenfunctions are Kummer (hypergeometric) functions in q(m), and the angular momentum eigenfunctions are unchanged. Under these circumstances the eigenfunctions are specified by a quantum number for each pseudorotation and for each puckering mode. The energy is independent of the rotation states, and is fixed entirely by the puckering amplitude(s). This occasionally a valid approximation, for rings containing hetero atoms especially.

V. Computer Algebra: Any Pedagogical Use?

So far all these examples have dealt with problems from research or graduate level theoretical courses. We have not made any argument that computer algebra can contribute to undergraduate level education. Experienced teachers have designed the physical chemistry course (and *a fortiori* the first year course) to minimize the mathematical demands so far as possible. This is certainly sound practice, since most students come to chemistry without much fluency and breadth of knowledge, skill, or interest in math. Speaking of the physical chemistry course, we must usually assume that although students have had calculus, somehow it just did not make much impact. The student's skills are rusty; calculus is not always used in his physics survey, and there would certainly have been no other call for this rather specialized talent. Further, however proficient he might be in the formal operations of the calculus, there is a major conceptual leap from those formal manipulations to describing phenomena in mathematical terms. Finally, the content of a year course in calculus is nowhere near sufficient to meet the demands of physical chemistry. Consider the grab-bag of mathematical topics one encounters in physical chemistry:

> partial differentiation
> line integration
> ordinary differential equations
> partial differential equations
> probability densities, averages, and variance
> combinatorics
> optimization with constraints

We can't expect any course except a custom departmental offering to meet our need to be acquainted (but only in the most superficial way, usually) with such a big bag of tricks. Such courses are not very common, but neither are they unheard of.

It is probably in the mathematical adjunct to physical chemistry that computer algebra can make some contribution to undergraduate training. It can shift the emphasis in such courses from mastering mechanical operation to <u>using</u> the tricks, without the frustration and distraction of mechanical errors. It can convey a sense of the power of the math, rather than disgust at one's own clumsiness. This is a result devoutly to be wished for, and diligently to be sought.

Thanks to Michael F. Brown (U. Va.) for helpful discussions.

Footnotes

[1] M. Boden, *Artificial Intelligence and Natural Man*, Basic Books, Inc. (new York, 1977).
[2] Quoted in "Computer Algebra", R. Pavelle, M. Rothstein, and J. Fitch, *Scientific American*, 136 (Dec 1981).
[3] B. Raphael, "The Thinking Computer: Mind Inside Matter" (W. H. Freeman, 1976); J. Bernstein, *The Analytical Engine: Computers Past, Present, and Future* (Wm. Morrow and Co., Inc., 1981, revised edition).
[4] J. Slagel's Symbolic Automatic Integrator, J. ACM, (1963) $\underline{10}$, 507.
[5] Proceedings, 1981 ACM Symposium on Symbolic and Algebraic Manipulation, P.S. Wang, ed. (Assoc. Comp. Mach., 1981).
[6] Contact A.C. Hearn, Computational Physics, University of Utah, Salt Lake City, Utah.
[7] Contact Dr. W. McCrae, Computer Center, U. Georgia, Athens, GA.
[8] Distributed by The Soft Warehouse, P.O. Box 11174, Honolulu, Hawaii 96828.
[9] A survey is given in [2].
[10] Quantum Chemistry Program Exchange, Chemistry Department, Indiana University, Bloomington, IN 47001.
[11] P. Winston and B. Horn, *LISP*, Addison-Wesley Publishing, Co. (Reading, MA, 1981), see BYTE, Aug 1979 (special LISP issue).
[12] J.H. Freed, G.V. Bruno, and C.F. Polnaszek, J. Phys. Chem. $\underline{75}$, 3385 (1971).
[13] Y. Siderer and Z. Luz, J. Magn. Res. $\underline{37}$ (1980).
[14] Several geometry programs are available from QCPE [10].
[15] J.A. Pople and D. Cremer, J. Am. Chem. Soc. $\underline{97}$, 1354 (1975).
[16] L.B. W. Jolley, *Summation of Series*, Dover Publ., Inc. (1961).

DON ZEBOLSKY

CHAPTER 5

Computer Applications in Chemistry

INTRODUCTION

    Creighton University is a Jesuit administered university located near downtown Omaha. There are around 2000 students in our College of Arts and Sciences and 200 chemistry majors. Forty graduate annually with a major in chemistry. Courses are offered outside the chemistry department in BASIC, FORTRAN and Pascal. Inside the department students use canned computer programs in analytical and physical chemistry. Computer Interfacing and Computer Applications in Chemistry are three semester hour electives. The latter course was designed during the summer of 1982. Five students enrolled in the fall semester of 1982. Prerequisites are physical chemistry and one course or its equivalent in computer programming.
    Creighton University has a UNIVAC 1100/60 mainframe with 40 time shared terminals about the campus. Four are located in the chemistry department. FORTRAN 77 and Danish Pascal, that version of Pascal implemented at the University of Copenhagen, are on the system. Four Apple II microcomputers are available in the department.
    Objectives of the course are to:

1) enable students to write their own programs in the high level language of their choice--Pascal, FORTRAN or BASIC;
2) comprehend and use selected numerical methods;
3) use the computer programs they write to solve problems and raise questions they might not raise without the computer.

The Maurits Escher print (1), "Lizards," is used to emphasize the third objective. Escher shows a cycle where a lizard crawls from the paper and comes to life with the aid of a book of science, a triangle and a dodecahedron. Following a puff, the lizard goes back into the sheet of paper. The cycle represents an iteration within a computer program. The mathematics helps bring the lizard to life. But it is each person's question and imagination that is paramount! Without that the rest is of no use.
    A brief outline of the course is shown in Table 1. After a week for a very simple introduction to the computer we spend three weeks on a review and comparison of the three high level languages, four on numerical methods, two on simulations and two on artificial intelligence. The text is "Computational Chemistry" by A. C. Norris (2).

TABLE 1. OUTLINE

COMPUTER BACKGROUND
REVIEW AND COMPARISONS OF PASCAL, FORTRAN AND BASIC
NUMERICAL METHODS
    CURVE FITTING, MATRICES, ROOTS AND INTEGRATION
    DIFFERENTIAL EQUATIONS AND TRANSFORMS
SIMULATIONS
ARTIFICIAL INTELLIGENCE

    A contract form of grading is used that is intended to enhance students' feelings that they are free to explore their own ideas on the computer. Students are asked to maintain a journal with all plans, ideas, lists and runs for weekly assignments and three projects. Two oral evaluations are scheduled to explain work and answer questions. If all work is complete with interpretations of individual questions for each program, I promise an A grade. If interpretations are lacking, a B grade may be earned by formally completing all work. Up to one-fourth of the work may be incomplete for a C grade.

COMPUTER BACKGROUND

    A survey of the hardware and software of our computer constitutes the first week of the course. Binary numbers are reviewed. Fixed and floating point storage for our computer is explained and roundoff and truncation errors discussed.

REVIEW AND COMPARISONS OF Pascal, FORTRAN AND BASIC

    A memory map is presented so students can distinguish a compiler from an interpreter and source from object programs. We then review all the features of elementary programming comparing each language side by side: input, output, variables, constants, loops, decisions, arrays, alphamerics, functions, subroutines and files. This review is done in the context of a problem that we discuss and work together in class. I wanted the problem to be simple in order to concentrate on the elements of programming, yet capable of development. Chosen were the amino acid separations of countercurrent distribution published in the original Craig machine article (3). Partition constants for five of the ten amino acids reported are used:

tryptophan, 2.0; phenylalanine, 0.99; tyrosine, 0.583; methionine, 0.397; and alanine, 0.135. Aqueous hydrochloric acid solutions are to be extracted with n-butanol. Our goal is to find the number of extractions needed to separate the amino acids. To begin with, the question is pushed back to an even simpler one: "How many extractions are needed to remove each amino acid from the aqueous phase?"

The program we start with is shown in its Pascal version in Table 2. This is compared with its BASIC

TABLE 2. PASCAL PROGRAM

```
PROGRAM SEPARATE
CONST
     INITIALCONC=1.5;
     KPARTITION=2;
VAR
     CONC,SUM,RATIO:REAL;
BEGIN
RATIO:=KPARTITION/(1+KPARTITION);
SUM:=0;
WHILE(INITIALCONC-SUM)>.01 DO
     BEGIN
     CONC:=(INITIALCONC-SUM)*RATIO;
     SUM:=SUM+CONC;
     WRITELN(CONC,SUM);
     END
END.
```

and its FORTRAN versions in a handout to the students. The required structure of Pascal is apparent even for this simple illustration. Note the declaration of constants and variables to begin with. Functions and subroutines would follow if there were any. Then the main is announced with BEGIN, and the loop is complete without GOTOs. The improvement in variable names from BASIC to FORTRAN to Pascal is discussed. At this time we review the Pascal loop options REPEAT...UNTIL and FOR... DO. All these loop options together nearly eliminate the need for GOTO commands. In the same fashion we review formatted inputs and outputs, arrays, string variables, many line branches, functions, subroutines and files. We use the same problem and extend the initial program. I ask each student to follow this process not only with the program we develop in class but also with a program they write for some analagous problem that requires iteration, such as isotope separation by gas diffusion, fractional distillation or another of their choice.

A feature unique to Pascal is the RECORD. Notice in Table 3 how the RECORD lets us naturally and readably group together aminoacids' names and their relevant

TABLE 3. RECORD IN PASCAL

```
CONST
     L=13;
TYPE
     STRING=PACKED ARRAY(1..L) OF ASCII;
     AATYPE=RECORD
          NAME:STRING;
          KPARTITION:REAL;
          EXTRACTIONS:INTEGER;
          ESSENTIAL:BOOLEAN;
     END;
VAR
     AMINOACID:ARRAY(1..5) OF AATYPE;
BEGIN
     WITH AMINOACID(I) DO ........
          IF NAME IN ESSENTIALAA THEN
          ESSENTIAL:=TRUE;
```

properties as if in a single array. The packed array is Pascal's way of storing a string variable and is more clumsy than strings in FORTRAN or BASIC. However the RECORD lets us associate real (the value of the partition constant), integer (number of extractions) and Boolean variables with the string. We could add other properties of interest to the RECORD such as pK values. Separate arrays would have to be used in FORTRAN or BASIC. We can define AMINOACID as an array of RECORDs and perform operations with each variable within each RECORD through the WITH...DO command. Pascal also uses SETs. Elsewhere in the program we can define ESSENTIALAA as the set of essential amino acids and with a single statement compare NAME with the SET: "If NAME in ESSENTIALAA then ESSENTIAL:=TRUE."

It is with RECORDs and SETs that Pascal comes into its own. Pascal is as easy to learn as BASIC, easier to read, and as powerful, fast and efficient as FORTRAN. RECORDs and SETs make Pascal special. I believe Pascal will become the high level language of the future in chemistry. About its only disadvantage is the lack of double precision.

Finally in this section we plot the results of the extractions, introduce the binomial distribution and compare with the normal Gaussian distribution. The Gaussian distribution arises naturally as one solution as the factorials of the binomial distribution rapidly become too large for the computer when the number of extractions are increased. We then question in class how to solve the original problem of separation. Students decide and each takes a programming assignment. At last in class we compare all results together in one plot. Students discover that two of the amino acids remain inseparable even after hundreds of extractions. This process of questioning and deciding in class is patterned after Dr. Charles Wales' GUIDED DESIGN method of instruction (4).

NUMERICAL METHODS: CURVE FITTING, MATRICES, ROOTS AND INTEGRATION

The problem I have chosen for the first Numerical Methods section follows.

> "Hydrogen is a possible replacement for methane but is it safe? It has a negative Joule-Thomson coefficient near room P and T. Use PV/RT,P data to find equations of state and deviations from ideality to try to answer."

In response I review the meaning of a hydrogen economy and of a negative Joule-Thomson coefficient. Then compressibility factor data (5) such as in Figure 1 is made available to the students. I have randomized the data to make its scatter obvious. Data sets at several temperatures are supplied on public data files. Students are asked to access data from these files and to store their results on files. By questions and discussion we recognize that a single equation for each gas would be useful. Discussion of least squares is followed by a matrix representation of the normal equations. Then we develop an algorithm in matrix notation that is identical regardless of the degree of polynomial fit.

Natural questioning before we try to judge the safety of hydrgen gas will lead us to want to predict the roots of our polynomial and perhaps the Boyle temperature, and to identify the fugacity coefficients by numerical integration over the data. We look at approximation, bisection and Newton's method to solve for polynomial roots and the Trapezoid and Simpson's

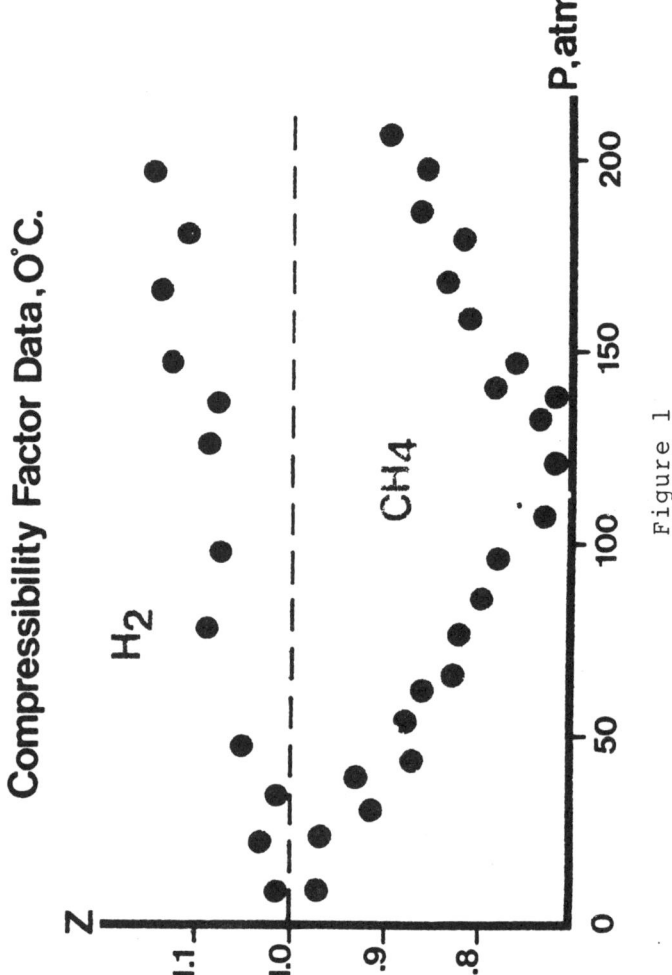

Figure 1

rules and the Romberg improvement for integration. As part of each class we discuss homework assignments that develop from previous classes as we work to write computer programs for our problem. Additional examples are provided; such as, heat capacity data, virial equations for colligative properties in solution, and spectral relationships and energy levels. Students are asked to choose one or fabricate another for themselves and write programs that use each numerical method.

NUMERICAL METHODS: DIFFERENTIAL EQUATIONS AND TRANSFORMS

The problem used to discuss the next section on differential equations and transforms is introduced by a demonstration of the Belousov-Zhabotinsky reaction. The recipe is taken from a book by James Espenson (6):

> "Add to a one liter beaker 600 cc of water, 60 cc of concentrated sulfuric acid, 20 grams of malonic acid, 7.8 grams of potassium bromate, 0.8 gram ceric salt and ferroin indicator."

Pipetting aliquots into a test tube leads to spatial striations that are faint but striking. Figure 2 shows a sketch of a photograph appearing in Moore's physical chemistry text (7). The actual blue and red color effect is more faint than appears in the sketch, yet it is clearly observable. The mechanism as reported by Field and Noyes (8) involves oscillation between two steady states. It is more complex than I want to begin with in class, yet the demonstration allows me to introduce simpler mechanisms that require numerical solution:

1) A ⟶ B ⟶ C

2) A + B ⟶ C ⟶ D

3) A + B ⇌ C ⟶ D

Mechanism one introduces the steady state. Mechanism two requires numerical solution of one differential equation, and mechanism three requires simultaneous solution of two differential equations. Taylor series and the Euler and the Runge-Kutta methods for solving differential equations numerically are compared.

The logarithm is used to explain the meaning of a transform. The integrated rate equation of a first

Belousov-Zhabotinsky Reaction Demonstration

"Physical Chemistry", Walter Moore, Prentice-Hall, 4th Ed., p.355.

Figure 2

order one step mechanism is used for illustration. La-Place and Fourier transforms are then compared. Some uses of Fourier transforms are mentioned, but LaPlace transforms are used in class. LaPlace transforms change differential equations into polynomials. If the differential equation is first order then the transform is linear. Transforms of two functions useful in kinetics are shown in Table 4. They can be derived with substitution and integration by parts. Table 5 illustrates

TABLE 4. LAPLACE TRANSFORMS

$$L = \int_0^\infty f(t)e^{-st}\, dt$$

$f(t)$	L
e^{kt}	$1/(s-k)$
df/dt	$sL(f) - f(0)$

TABLE 5. TRANSFORMATION PROCEDURE
FOR A TRIVIAL CASE

$$A \xrightarrow{k} B \qquad dA/dt = -kA$$

TRANSFORMED RATE EQUATION

$$sL(a) - A(0) = -kL(A)$$

$$L(A) = A(0)/(s+k)$$

INVERSION

$$A = A(0)e^{-kt}$$

the procedure for the trivial case of one first order reaction step. The transformed rate equation, linear in $L(A)$, can be solved for $L(A)$, which upon inversion yields the expected integrated rate equation. LaPlace transforms can be used in this way to solve any mechanism of first order reactions (9). Mechanisms with a second order reaction step are more difficult and require a numerical iteration technique (10).

SIMULATIONS

The use of deterministic models to simulate data is mentioned briefly. Examples are the van der Waals model for compressibility curves and model NMR spectra. Then I give students a random number generator following the suggestions of Forsyth (11), modified for the UNIVAC 1100's 36 bit word, and introduce the Monte Carlo technique. The Monte Carlo technique is beautiful and widely applicable. My favorite applications to education in the literature are listed below as I give them to my students:

1) Maxwell-Boltzmann distribution (Schettler [12]);
2) Brownian Motion (Soltzberg [13]);
3) Consecutive Reaction Rates (Rabinovitch [14]).

We study the reaction rate simulation in class because it fits the previous section so nicely.

A useful procedure for the Monte Carlo technique follows. In a loop within the program a set of random numbers are generated. The random numbers are then scaled or transformed before testing. If the test condition is fulfilled then certain consequences occur depending on the simulation. The grid in Figure 3 illustrates the procedure using Rabinovitch's approach (14). The grid represents a two or three dimensional array that simulates the reaction vessel. A and B represent reactant molecules and blanks the solvent. A set of random numbers, ten let's say, are generated. Grouped in pairs the five pairs are scaled to span the array. Each pair after scaling represents a location in the vessel. Suppose $k_1/k_2 = 3/2$. Then the first three pairs refer to the first step. If an A is found at a location it is changed to B. The next two pairs of the set of five is tested for the second step. The location of a B would be changed to a C. The cycle is repeated hundreds of times. The number of A, B and C molecules are then printed on a graph versus trial number as time to complete the simulation. This project is among the most satisfying of the course. Class members agreed upon a flow chart for the A ⟶ B ⟶ C mechanism. Each worked separately on their own subroutine between classes. We created a public file and combined the subroutines into one program during the next class. Some editing was needed to get the program to run error free and to show the output the way we wanted. Rate constants and initial concentrations were then varied empirically and concentration vs. time

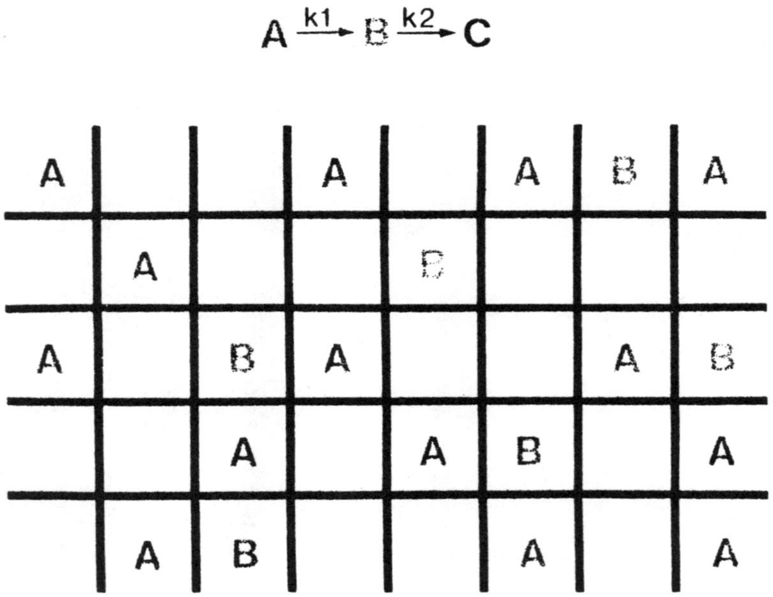

$$A \xrightarrow{k1} B \xrightarrow{k2} C$$

Figure 3

plots were examined for insight into the steady state.

ARTIFICIAL INTELLIGENCE

A brief introduction to the subject of artificial intelligence completes the course. A search program for organic synthesis coded for twenty compounds and thirty reactions from Isenhour and Jurs (15) is studied. A seminar was given about the expert consultant system COMMES operative in Creighton University's schools of health sciences and available throughout Nebraska (16). Utility functions are defined and we look at some simple decision matrices from non-zero sum game theory (17). Computer learning is questioned with the game of Hexapawn (18) illustrated in Figure 4. Hexapawn is played on a nine square board. Moves are identical to that of a pawn in chess. The player starts. Then the computer takes a turn. Three turns for each ends the game. All positions and all moves can be put into a computer. Moves and positions are randomly chosen by the computer. The computer learns by eliminating moves that cause it to lose. Finally pattern vectors, decision vectors and decision planes for binary classifiers from pattern recognition are discussed, and we study a flow sheet of Isenhour and Jurs' program "The Learning Machine" (15). We concentrate on their subroutine LEARN in order to understand how the computer learns from the training set.

CONCLUSION

The course closes as it began with another print of Escher's (1), "Reflections." It is my hope with this class to stimulate students to confidence that they can use computer programming to view problems in new ways as Escher does with the reflections of trees in a mud puddle.

ACKNOWLEDGEMENTS

Appreciation is extended to Creighton University for a 1982 Summer Faculty Development Grant, to the members of the Computers in Chemistry Task Force of the American Chemical Society who offered workshop opportunities that enhanced my experience in computer programming, and to Tim Ream, Randy Pritza, John Espinosa, Doug Leonovicz and Miki Nobuyuki, members of this first class.

THE GAME OF HEXAPAWN

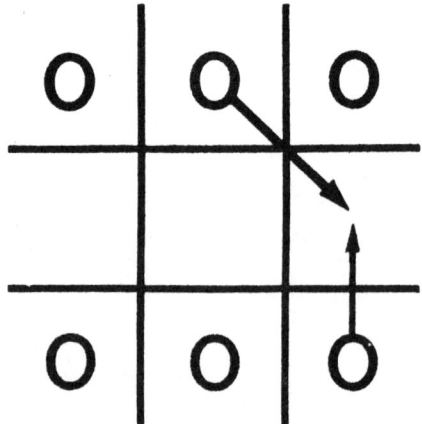

Martin Gardner, "Game Learning Machine", Scientific American 1962, 138 (March).

Figure 4

REFERENCES CITED

1. Escher, M. C., "The Graphic Work of M. C. Escher," translated by John E. Brigham, Meredith Press, NY, 1967; "The World of M. C. Escher," edited by J. L. Locher, Abrams, Inc., NY, 1971.
2. Norris, A. C., "Computational Chemistry," Wiley, NY, 1981.
3. Craig, L. C., W. Hausmann, E. H. Ahrens, Jr., and E. J. Harfenist, Analytical Chemistry 23, 1236, 1951.
4. Wales, Charles E. and Robert A. Stager, "Educational Systems Design," Morgantown, West Virgnia, 1974.
5. Washburn, Edward W., editor, "International Critical Tables," McGraw-Hill, NY, 1928.
6. Espenson, James H., "Chemical Kinetics and Reaction Mechanisms," McGraw-Hill, NY, 1981, p. 143.
7. Moore, Walter J., "Physical Chemistry," 4th ed., Prentice-Hall, Englewood Cliffs, 1972, p. 355.
8. Field, Richard J. and Richard M. Noyes, Journal of Chemical Physics 60, 1877 (1974).
9. McLaughlin, Eileen and R. W. Rozett, Journal of Chemical Education 49, 482 (1972).
10. Bellman, R. E., Robert E. Kalaba and JoAnn Lockett, "Numerical Inversion of the LaPlace Transform," Elsevier, NY, 1966, p. 121.
11. Forsythe, George E., Michael A. Malcolm and Cleve B. Moler, "Computer Methods for Mathematical Computations," Prentice-Hall, 1977, p. 242.
12. Schettler, Paul, Journal of Chemical Education 51, 250 (1974).
13. Soltzberg, Leonard, Arvind A. Shah, John C. Saber and Edgar T. Canty, "BASIC and Chemistry," Houghton-Mifflin, Boston, 1975, p. 145.
14. Rabinovitch, B., Journal of Chemical Education 46, 262 (1969).
15. Isenhour, Thomas L. and Peter C. Jurs, "Introduction to Computer Programming for Chemists: FORTRAN," 2nd ed., Allyn-Bacon, Boston 1979.
16. Evans, Steven, "The Total-System Design of Instruction: A Project Overview," Creighton University, Omaha, 1981.
17. Rubinstein, Moshe F., "Patterns of Problem Solving," Prentice-Hall, Englewood Cliffs, 1975.
18. Gardner, Martin, Scientific American 1962, 138.

Vector and Scalar Computers:
A Comparison

I. Introduction

In this paper we will examine a number of concepts that may be of interest to any individual who is either contemplating the use of a supercomputer for problem solving or is currently using one. The motivation to examine these concepts is largely economic. Since supercomputers are very expensive, it is desirable to develop programming and computational methods that will maximize the available resource. This in turn may enable the user to solve new and more complex problems; at the very least, it should enable him to to determine more accurate solutions to existing problems. With the advent of micro-miniaturation, the cost of computing is rapidly decreasing. Indeed, it has been stated [3] that by 1990 a computer with the power of a current supercomputer could be available on a chip that would sell for 200 dollars. Thus, the concepts that are currently considered relevant only to supercomputing may soon become relevant to all computing. The subjects that will be discussed are independent of any particular discipline. They are just as important in chemistry as in atmospheric science. Additional introductory material on supercomputing is contained in [2].

The paper is divided into two major parts: Sections I and II are concerned with architectural factors that must be considered in order to make the most efficient use of a supercomputer. Sections III, IV, and V are concerned with algorithmic considerations. We show, first by simple example, how to redesign algorithms so that they make efficient use of the special architectural features of the supercomputer. We then analyze two specific computations namely, the solution of tridiagonal systems of equations and the fast Fourier transform. These two algorithms are fundamental to scientific computing and consume a considerable amount of supercomputing resource. In this paper we attempt to separate real from imaginary concerns, or practical from theoretical considerations. Nevertheless, the theoretical considerations are presented, since they will soon become the practical considerations of using possibly the next generation of supercomputer.

Roughly speaking the current supercomputers can be placed into two categories namely, pipelined or parallel. The first category, pipelined supercomputers, gain speed by using an assembly-line approach. In order to take advantage of the assembly line, or pipeline, the user must have many of the same type of calculations. A good example is the addition of two sequences or vectors, i.e. given the vectors a_i and b_i for $i = 1, \ldots, n$, then compute a third vector $c_i = a_i + b_i$. In the pipeline computer the adder is highly segmented, with each segment working on a particular part of the addition $a_i + b_i$. Once it completes that part for $a_i + b_i$, it can then begin on the same part for $a_{i+1} + b_{i+1}$. Hence, several c_i's are in the process of being computed at the same time. Once all segments are busy, or the pipeline is full, the c_i's come off the assembly line at a very rapid rate. Each c_i exits the pipe every small fraction of a second, which is called the basic cycle time of the computer.

If s is the time required to fill the pipe and Δt_ν is the basic cycle time, then the total time required to compute the c_i's is $t_\nu = s + n \Delta t_\nu$. It is important to note that the pipeline computer is in fact a sequential computer, since the results are computed one after the other. The only difference between a pipeline and a sequential computer is that for a certain set of calculations, like the one above, the pipeline computer is able to compute the results much faster. For these "vector" calculations the pipeline computer is a very efficient sequential computer. A pipeline computer is also called a vector computer, since it is primarily used for vector computations Examples of pipeline supercomputers include the CRAY-1, CRAY X-MP, and CYBER 205.

The second category of supercomputer is the parallel computer. Examples include the Illiac IV, International Computers Ltd. Distributed Array Processor (DAP), and the Homogeneous Element Processor (HEP) by Denelcor. At the current state of the art the overall performance of a parallel computer is not that much different from that of a pipeline computer, even though there is a fundamental difference in their design. The parallel computer computes all the c_i's at exactly the same time. The parallel computer consists of n processors, each of which is capable of peforming an instruction set, including addition, multiplication, etc. Each processor contains one each of the a_i and b_i, and on command all the c_i's are computed simultaneously in a time Δt_p that is independent of n if n processors are available.

Based on these considerations alone, the parallel architecture would seem to be clearly preferred over the pipeline architecture. However, the pipeline architecture is actually economically competitive in the sense that it produces about the same number of computations per dollar as the parallel computer. The multiprocessor parallel computer is more expensive than the pipeline computer unless a cheaper technology is used. However, the cheaper technology results in a larger Δt_p, which can be comparable to $s+n\Delta t_v$ when n is about equal to the number of processors in existing parallel computers.

The current trend is to combine the features of pipeline and parallel architectures. For example, the CRAY X-MP includes two pipes, each of which can be viewed as a separate processor. Thus, it can be viewed as a parallel computer, but with only two processors. The CYBER 205 has a proposed four pipes and therefore also fits into this category.

The performance of the current generation of supercomputers is the result of continued advances in the following three major architectural areas:

1. New technologies combined with miniaturization, which have contributed to a steady decrease in cycle time.

Vector and Scalar Computers: A Comparison

2. The use of pipeline or vector architectures, in which the computations are still performed sequentially but in a very efficient assembly-line manner.

3. As a result of reduced costs, recent trends toward multiprocessors or multiple pipelines.

If these trends continue, future computers will probably incorporate advances in each of these areas. That is, we can expect multiple-pipelined processors in a variety of configurations - even on chips - with ever-decreasing cycle times. The current trend toward multiprocessors suggests that it is now cheaper to double the performance of a computer by doubling the number of processors rather than by halving the basic cycle time. If current supercomputers are available on a chip in 1990, then it is difficult to imagine a supercomputer without a large number of processors. However, difficult problems of communication (interconnection) and orchestration (control) must be solved before the efficient use of many processors can be guaranteed.

II. Architectural Considerations

Performance measures that have been used for past generations of computers are inadequate when they are applied to supercomputers. As for earlier computers, one performance measure of a supercomputer is its basic cycle time, or the number of operations that it can perform per second, or more commonly, millions of floating point operations per second (megaflops). However, the rates that are given are optimum and apply only to a very small class of problems. The actual performance may be one or two orders of magnitude below the rated megaflops. Nevertheless, if only a single performance measure is used, the optimum rate is probably as reasonable a choice as any other. A number of other factors influence the performance of a supercomputer. Each entry in this list is discussed below.

1. Basic cycle time
2. Scalar vs vector mode
3. Number of pipelines or processors
4. Vector length
5. Word length
6. Start-up time
7. Chaining
8. FORTRAN compiler

       9. Operation mix
    10. Memory-to-memory vs in-register
    11. Contiguous vector elements
    12. The algorithm

1. **Basic cycle time**. The basic cycle time is of course still important for overall performance. Therefore, it is an important, but not all-encompassing, measure of performance; other factors must be considered before an accurate assessment can be made.

2. **Scalar vs vector mode**. Not all computations are of the type described in the preceding section. For short vectors, particularly those with just a single element, the pipeline or parallel architecture is not efficient. Nevertheless, the computer must be able to handle these computations in what is called its scalar mode. When computing in scalar mode the megaflop rate is significantly diminished, possibly by as much as a factor of 10.

3. **Number of pipelines or processors**. Current trends are to multiple-pipe computers, of which the CYBER 205 and CRAY X-MP are examples. A two- or four-pipe computer can have about two to four times the capability of a single-pipe computer, depending on the size of memory and the way that the pipes and memory are interconnected. The performance also depends on the particular application and on whether the computation can be distributed across the pipes.

4. **Vector length**. As discussed in the first section, the total time for a vector calculation is $t = s + n \Delta t_V$. The start-up time s can be substantially larger than Δt_V. Therefore, the megaflop rate increases with the vector length n. An interesting parameter $n_{1/2}$ that has been introduced by Hockney [1] is the length of a vector required to achieve half the maximum performance. Hockney gives $n_{1/2} = 10$ for the CRAY-1 and $n_{1/2} = 100$ for the CYBER 205. Unfortunately, $n_{1/2}$ can vary, even on the same computer, depending on some of the other factors in this list.

5. **Word length**. The megaflop rate on the CYBER 205 can be doubled by computing with 32-bit words rather than 64-bit words.

6. **Start-up time**. If the time s required to load the pipe is excessive, then n must be very large in order to achieve near-maximum megaflop rates. This is a significant factor since n is often restricted by memory or computing time considerations. The parameter $n_{1/2}$ is a measure of the significance of the start-up time.

7. **Chaining**. If more than one operation, say addition and multiplication, is performed on the assembly line, or in the pipe, then the operations are said to be chained. Chained operations can significantly increase performance. Any additional operations, following the first are essentially free, since they are computed in a time proportional to $\Delta t_{1/}$, which is independent of the vector length n. This changes the way in which the efficiency of an algorithm should be measured. Up to the present, the performance of an algorithm has been measured by the number of arithmetic operations. With chaining, some of the operations are no longer apparent and the efficiency of the algorithm becomes machine-dependent Indeed, for some calculations a reasonable measure of efficiency is the number of DO loops rather than the number of operations.

8. **FORTRAN compiler**. The performance of the FORTRAN compiler is a very significant factor, particularly for a new user. To a very great extent, it determines the ease of transition to a supercomputer and the efficiency of the initial use. Transitions are always difficult, but a good compiler can considerably ease them. It is particularly significant to the supercomputer user when the compiler must also, where possible, make use of vector or parallel architecture.

There is currently a vast difference in the degree of compiler sophistication. The less sophisticated compilers require nonstandard FORTRAN in order to make efficient use of the supercomputer architecture. Although these programs are difficult to read and port, they can use the computer very efficiently. The more sophisticated compilers are able to recognize most situations in which vector code can be compiled. None of the compilers can recognize all situations in which vector code can be compiled, however, and thus machine language programming is still more efficient - but also more expensive. There are, of course, situations in which the algorithm must be changed in order to make efficient use of the architecture. Several such cases are discussed in the following sections.

9. <u>Operation mix</u>. Depending on the types of operations and how they can be sequenced, the compiler or programmer can take advantage of architectural features, such as by chaining. Although the chaining feature is usually associated with the CRAY-1, these considerations also apply to the CYBER 205. The megaflop rate can vary by as much as a factor of two, depending on whether a "dyad" or "triad" is computed. A dyad is a computation of the form that was discussed above, namely, computing $c_i = a_i + b_i$, whereas a triad is of the form $c_i = a_i + cb_i$, where c is a constant. On the CYBER 205 the megaflop rate for a triad is double that for a dyad.

10. <u>Memory-to-memory vs in-register computation</u>. On the CRAY-1 the data are first moved from memory to vector registers. The computation then proceeds to and from these registers at a very high rate, up to 160 megaflops. This is referred to as the in-register rate. However, eventually the computation requires access to memory. When the time required to access and write memory is included, then the computation is referred to as memory-to-memory. In practice, memory-to-memory computations are made at about half the rate of in-register computations. Memory-to-memory computations are more efficient on the CRAY X-MP than they are on the CRAY-1, since the X-MP has more paths or ports into memory.

11. <u>Contiguous vector elements</u>. The speed of all supercomputers depends on how the elements of a vector are stored. The most efficient configuration is storage in contiguous memory locations in which case the elements are said to have stride 1. If the elements do not have stride 1, then the performance of the computer can be substantially reduced. On the CYBER 205, the megaflop rate can be reduced by a factor of 10. Performance of the CRAY-1 is also impaired, but not to the extent of the CYBER 205. The CRAY-1 performance degrades gradually as the stride increases as a power of two, i.e., as the stride increases as a power of two from two to 16.

12. <u>The algorithm</u>. The final factor, and perhaps the most important, is the selection of an algorithm

Vector and Scalar Computers: A Comparison 129

for a particular computation. Most computations can be performed in a number of different ways called algorithms. Although the same result is obtained, the intermediate computations can be very different. The idea is to select an algorithm such that the intermediate calculations can be performed efficiently on a supercomputer. Algorithms in which none or very few calculations can be performed at the same time are called scalar algorithms; these contrast with parallel or vector algorithms, in which many calculations can be performed at the same time. Computational mathematicians have been concerned for some time with developing parallel algorithms for traditionally scalar calculations in anticipation of the supercomputers that are now and yet to become available. The parallel algorithm is important because it is the only factor for which the savings depend on the vector length n rather than some constant factor. In the next section we will give a simple example that illustrates how the performance of a parallel computer can be increased by a factor of $n/\log_2 n$ by replacing a scalar algorithm with a parallel algorithm.

III. Algorithmic Considerations.

In this section we examine the adaptation of algorithms to the supercomputer. In most cases there are many ways to compute a particular result, and a very simple example of this will be presented in this section. The final result is always the same, but the intermediate computations can be very different. Thus there may be several different algorithms for computing the same result. The fast Fourier transform is a very efficient algorithm for computing the same result that is computed by the slow Fourier transform. Efficient algorithms for super computers are characterized by having many computations of the same kind. These algorithms are called vector or parallel algorithms, which are in contrast to scalar algorithms, in which none or very few calculations can be performed at the same time.
  We will present, by a simple example, the development of a parallel algorithm as an alternative to a given scalar algorithm. This development is independent of the manner in which a compiler produces vector code. We are examining the case in which the basic algorithm is scalar, so that a compiler, no matter how ingenious, could not produce vector code. The example will demonstrate that a parallel alternative to a scalar algorithm cannot be automated. The parallel algorithm for solving tridiagonal systems, which is

outlined in the next section, is more representative of the difficulty encountered in the development of actual parallel algorithms.

Consider now the computation and tabulation of the partial sums of the sequence y_k. That is, given the sequence y_k, we wish to compute

$$s_j = \sum_{k=1}^{j} y_k \qquad j = 2,\ldots,n.$$

The scalar algorithm for computing s_j starts by setting $s_1 = y_1$ and then computing the rest of the s_j using the recurrence relation

$$s_j = s_{j-1} + y_j \qquad j=2,\ldots,n$$

This computation is a classic example of a scalar algorithm, since none of the additions can be done simultaneously. For example s_{10} cannot be computed until s_9, which cannot be computed until s_8, and so forth.

In order to develop a parallel algorithm, the following notation is introduced

$$s_{i,j} = \sum_{k=i}^{j} y_k$$

Thus, $s_{i,j}$ is just the sum of the y_k with subscripts k that satisfy $i \leq k \leq j$.

Using this notation, the problem becomes one of computing $s_{1,j}$ for $j=1,\ldots,n$. To this end we present an algorithm by Stone [4]. The presentation is simplified if we select a fixed value of $n = 8$. In each of the steps below, the calculations can be performed simultaneously. We begin with Phase I (reduction)

step 1

$$s_{1,2} = y_1 + y_2 \ ; \ s_{3,4} = y_3 + y_4 \ ; \ s_{5,6} = y_5 + y_6 \ ; \ s_{7,8} = y_7 + y_8$$

step 2

$$s_{1,4} = s_{1,2} + s_{3,4} \ ; \ s_{5,8} = s_{5,6} + s_{7,8}$$

step 3

$$s_{1,8} = s_{1,4} + s_{1,8}$$

At this point we have a few of the desired partial sums, namely, $s_{1,2}$, $s_{1,4}$, and $s_{1,8}$. The remainder are computed in

Phase II (backsubstitution)

step 1

$$s_{1,6} = s_{1,4} + s_{5,6}$$

step 2

$$s_{1,3} = s_{1,2} + y_3 \; ; \; s_{1,5} = s_{1,4} + y_5 \; ; \; s_{1,7} = s_{1,6} + y_7$$

At this point all of the partial sums $s_{1,j} = s_j$ for $j = 1,\ldots,8$ have been computed. On a parallel computer with at least four processors the calculations in each step can be performed simultaneously. Therefore the total time required by the reduction phase is $3\Delta t_a$, where Δt_a is the time required for an addition. More generally if $n=2^m$, then the time required for the reduction phase would be $m\Delta t_a$ or $\log_2 n \Delta t_a$. The time required for the backsubstitution phase is $(\log_2 n - 1)\Delta t_a$. Thus, the total time is $(2\log_2 n - 1)\Delta t_a$. It is customary to state simply that the time is proportional to $\log n$ or that the algorithm is $O(\log n)$. On the other hand, the time would be proportional to n if we were to use the scalar algorithm, even on a parallel computer. Thus, the speed-up factor is $O(n/\log n)$. This corresponds to a substantial speed-up - the same as one obtains by using the fast Fourier transform instead of the slow Fourier transform.

If $n=1024=2^{10}$, then theoretically the parallel algorithm should be 1024/10 or approximately 100 times faster than the scalar algorithm. However in practice on the CRAY-1, the parallel algorithm computes only about five times faster than the scalar algorithm. Clearly a factor of 5 can result in a considerable saving of computing resource. Nevertheless, a natural question arises: Why is there such a difference between theory and practice? There are four reasons, which are basically all related to the fact that the CRAY-1 is not a parallel computer.

1. As mentioned in the first section, the CRAY-1 is in fact a very efficient sequential computer, and therefore the computing time must be proportional to n rather than $\log n$.

2. The stride or distance between elements in storage increases as powers of two as the steps progress. In step 2, the data are accessed with stride 2; at step 3, it would be 4; at step 5 it would be 8; and so forth. But in the first section we observed that the performance of any current supercomputer degrades if the data are not stored in contiguous locations. In particular, the performance of the CRAY-1 degrades if the stride is a power of two.

3. The vectors get shorter from one step to the next. That is, the number of computations that can be done at the same time decreases as the steps progress. Therefore, the start-up time becomes more of a factor. Observe that although the amount of computation in step k+1 is half that of step k (in phase I), the computing time in step k+1 will not be half that of step k.

4. The parallel algorithm requires twice as many actual additions as the scalar algorithm. On the CRAY-1, each addition makes a contribution to the total time.

Clearly, we would not expect to achieve the theoretical speed-up on the CRAY-1. However, even if the parallel algorithm were to be run on any of the current generation of supercomputers, the theoretical performance would not be obtained. The reason is that the processors are only locally connected. That is, only contiguous processors are connected, whereas the data flow for the parallel algorithm follows a binary tree path in which the ith processor should be connected to the $i+2^j$th processor for $j = 0,1, \ldots$. In timing the parallel algorithm we ignored the time required to transmit the data between processors. This is reasonable only if the processors are interconnected in such a way as to support the flow of data as dictated by the algorithm. Since none of the processors are interconnected in a tree fashion, the time required for the parallel algorithm would be proportional to n. Even on a parallel computer!

In this section we have presented the development of a simple parallel algorithm that can be useful on the current generation of supercomputer, but that will have to wait (but not for long) for subsequent generations of supercomputer to realize its full potential. The algorithm that is presented in the next section fits into the same class; that is, it is useful now but its full potential has not yet been realized.

IV. Solving Tridiagonal Systems.

In this section we will outline a parallel algorithm for solving tridiagonal linear systems of equations. These systems occur frequently in scientific applications; in particular, they must be solved repeatedly in order to solve the ordinary and partial differential equations that model geophysical processes. Since a great deal of computing resource is used to solve tridiagonal systems, it is reasonable to invest some effort in making that computation as efficient as possible. Several methods for solving tridiagonal systems on supercomputers are given in [6], including a method [5] that will be outlined below. Note that in this paper, as in most computational literature, we examine only those computations for which the development of a parallel algorithm is nontrivial. If many, say m, are to be solved simultaneously, then the development of a parallel algorithm is straightforward, since there are at least m of the same kind of calculations at each step. In fact, this turns out to be the best approach when many systems are available for solution simultaneously and memory is plentiful. Nevertheless, the parallel algorithms are also useful for single systems and will significantly outperform all other algorithms on future computers, which may have thousands of processors.

Until the advent of vector and parallel computers, one of the most satisfactory algorithms in computational mathematics was the Gauss algorithm for solving tridiagonal systems. It is very fast, with an operation count that is proportional to n, compared with n^3 for a full linear system. It also requires very little memory and has good error characteristics.

Given the nxn matrix

$$A = \begin{vmatrix} b_1 & c_1 & & & & & \\ a_2 & b_2 & c_2 & & & & \\ & a_3 & \cdot & & & & \\ & & & \cdot & & & \\ & & & & \cdot & & \\ & & & & & c_{n-1} & \\ & & & & & a_n & b_n \end{vmatrix}$$

and right side $y^T=(y_1,y_2,\ldots y_n)$, then we wish to find a solution $x^T=(x_1,x_2,\ldots y_n)$ such that $Ax = y$. Gauss elimination for tridiagonal systems assumes a particularly simple form. If we define $u_1=b_1$, $w_1=y_1$ and for $i = 2, \ldots, n$ compute

$$u_i = b_i - \frac{a_i}{u_{i-1}} c_{i-1}$$

$$w_i = y_i - \frac{a_i}{u_{i-1}} w_{i-1}$$

then the solution x satisfies the upper bidiagonal system

$$\begin{vmatrix} u_1 & c_1 & & & \\ 0 & u_2 & c_2 & & \\ & & \cdot & & \\ & & & \cdot & \\ & & & & c_{n-1} \\ & & & & u_n \end{vmatrix} \begin{vmatrix} x_1 \\ x_2 \\ \cdot \\ \cdot \\ x_{n-1} \\ x_n \end{vmatrix} = \begin{vmatrix} w_1 \\ w_2 \\ \cdot \\ \cdot \\ w_{n-1} \\ w_n \end{vmatrix}$$

From the last equation $x_n = w_n/u_n$. The remaining elements x_i can be computed by backsubstitution for $i=n-1, n-2, \ldots, 1$

$$x_i = \frac{w_i - c_i x_{i+1}}{u_i}$$

This elegant form of the Gauss algorithm has been used for years to compute solutions to tridiagonal systems. However, it is somewhat flawed for use on a supercomputer. As with the scalar algorithm in Section III, virtually none of the computations can be done simultaneously; i.e., u_i and w_i cannot be computed until u_{i-1} and w_{i-1} have been computed. Also, x_i cannot be computed until x_{i-1} is computed. In addition, if $u_i=0$ for some i, then the algorithm fails. Usually this is not a serious problem, since it can be shown that $u_i \neq 0$ for most systems of interest. Nevertheless,

Vector and Scalar Computers: A Comparison 135

the parallel algorithm that will be outlined is defined for any nonsingular tridiagonal matrix. It is based on Cramer's rule, in which the only division is by the determinant of the matrix. Thus, the algorithm fails only if the matrix is singular.

Since the parallel algorithm is based on Cramer's rule, we define the following determinants,

$$d_i = \begin{vmatrix} b_1 & e_1 & & & & \\ a_2 & b_2 & c_2 & & & \\ & a_3 & \cdot & & & \\ & & & \cdot & & \\ & & & & \cdot & c_{i-1} \\ & & & & a_i & b_i \end{vmatrix}$$

$$g_i = -a_{i-1} \begin{vmatrix} b_i & c_i & & & & & \\ a_{i+1} & b_{i+1} & & c_{i+1} & & & \\ & a_{i+2} & \cdot & & & & \\ & & \cdot & & & & \\ & & & \cdot & & & c_{n-1} \\ & & & & & a_{n-1} & b_n \end{vmatrix}$$

$$s_i = \begin{vmatrix} b_1 & c_1 & & & & y_1 \\ a_2 & b_2 & c_2 & & & \cdot \\ & a_3 & \cdot & & & \cdot \\ & & & \cdot & & \cdot \\ & & & & \cdot & y_{i-1} \\ & & & & a_i & y_i \end{vmatrix}$$

$$u_i = \begin{vmatrix} y_i & c_i & & & & \\ y_{i+1} & b_{i+1} & c_{i+1} & & & \\ \cdot & a_{i+2} & \cdot & & & \\ \cdot & & \cdot & & & \\ \cdot & & & \cdot & & \\ & & & & c_{n-1} & \\ y_n & & & & a_n & b_n \end{vmatrix}$$

Starting with Cramer's rule, it has been shown [5] that the solution can be given in terms of the sequences d_i, g_i, s_i, u_i as:

$$x_i = \frac{d_{i-1} u_i + g_{i+1} s_i}{d_n}$$

The x_i can be computed simultaneously once the sequences d_i, u_i, g_i, and s_i have been determined. Therefore it remains to determine parallel algorithms for computing these sequences. A parallel algorithm will be developed for the sequence d_i only since the development for g_i, s_i, and u_i is either similar or simpler. By expanding the determinant d_i in terms of its ith row, we obtain the three-term recurrence

$$d_i = b_i d_{i-1} - a_i c_{i-1} d_{i-2}. \tag{1}$$

Although the determinants d_i can be computed from this recurrence, it is not a parallel algorithm since d_i cannot be computed until d_{i-1} is computed and so forth. However, a parallel algorithm can be developed in the following way:

Define

$$e_i = c_i d_{i-1} \tag{2}$$

then (1) and (2) can be written in 2x2 matrix notation.

$$\begin{bmatrix} d_i \\ e_i \end{bmatrix} = \begin{bmatrix} b_i & -a_i \\ c_i & 0 \end{bmatrix} \begin{bmatrix} d_{i-1} \\ e_{i-1} \end{bmatrix}$$

with solution

$$\begin{bmatrix} d_i \\ e_i \end{bmatrix} = \prod_{j=1}^{i} \begin{bmatrix} b_j & -a_j \\ c_j & 0 \end{bmatrix} \begin{bmatrix} 1 \\ 0 \end{bmatrix} \quad (3)$$

The parallel algorithm for computing the matrix partial products in (3) is almost the same as the parallel algorithm given in section III for computing the partial sums of a sequence y_i. The only difference is that the elements y_i are replaced by the 2x2 matrices

$$Q_i = \begin{bmatrix} b_i & -a_i \\ c_i & 0 \end{bmatrix}$$

and scalar addition is replaced by 2x2 matrix multiplication. Once the partial matrix products are known, then the sequence d_i is given as the upper left element in the partial product matrices. It turns out that the sequence g_i can also be computed from the matrix partial products. The parallel algorithm for computing u_i and s_i is somewhat simpler, and the details are given in [5].

With 5n processors, this algorithm requires $5\log_2 n$ multiplications and additions. The algorithm also requires 6n storage locations, compared with 5n for a Gauss elimination algorithm that does not fail if $u_i = 0$. Computing times are given below in Table 1.

TABLE 1

n	Gauss Elimination CRAY-1	Cramer's Rule CRAY-1	Cramer's Rule CYBER 205* (2 Pipe)
	Computing Times in Microseconds For Solving Tridiagonal Systems		
32	178	165	237
64	339	229	309
256	1323	515	563
1024	5258	1548	1303

*These times are courtesy of James H. Koehler, University of Minnesota

V. The Fast Fourier Transform.

In most respects, the FFT is a parallel algorithm. On a parallel computer with n processors, it is easy to envision how to connect the processors and how to distribute the computation so that the computing time would be reduced by a factor of n when compared with the time required by the FFT on a single processor. The situation is somewhat different on a pipeline or vector computer, and this difference will be the focus of this section.

A parallel algorithm is also usually an efficient algorithm for a pipeline or vector computer. However, in the case of the FFT, the algorithm for a vector computer differs somewhat from the algorithm for a parallel computer. Therefore the algorithmic design considerations for a parallel and a vector computer are not always identical. In this section we will outline the development of a vector algorithm for the FFT. The details are given in [7].

In order to understand the vector algorithm it is first necessary to understand the FFT algorithm itself. We will describe a very simple version of the FFT, which is not satisfactory for computational purposes but is useful for expository purposes. A software package called FFTPACK, which is more representative of the state of the art, is available from the Scientific Computing Division, NCAR, P.O. Box 3000, Boulder, CO 80307. The package consists of about 3,000 lines of FORTRAN code for computing a number of fast Fourier transforms.

Given a sequence y_j, then the discrete complex Fourier transform c_k is given by

$$c_k = \frac{1}{n} \sum_{j=1}^{n} y_j e^{-ijk\frac{2\pi}{n}} \qquad k=1,\ldots,n \qquad (4)$$

with inverse

$$y_j = \sum_{k=1}^{n} c_k e^{ijk\frac{2\pi}{n}} \qquad j=1,\ldots,n$$

To compute c_k using (4) requires $n(n-1)$ complex additions and $n(n+1)$ complex multiplications. We show now that if n is divisible by two, then the c_k can be

Vector and Scalar Computers: A Comparison

computed with about half the number of operations using a simple modification of (4), which is called the splitting algorithm. If the even and odd terms are separated in the sum on the right side of (4), then

$$c_k = \frac{1}{n}\sum_{j=1}^{n/2} y_{2j} e^{-i2jk\frac{2\pi}{n}} + \frac{1}{n}\sum_{j=1}^{n/2} y_{2j-1} e^{-i(2j-1)k\frac{2\pi}{n}} \quad (5)$$

or

$$c_k = \frac{1}{n}\sum_{j=1}^{n/2} y_{2j} e^{-ijk\frac{2\pi}{n/2}} + \frac{1}{n} e^{ik\frac{2\pi}{n}} \sum_{j=1}^{n/2} y_{2j-1} e^{-ijk\frac{2\pi}{n/2}} \quad (6)$$

If we define

$$a_k = \frac{2}{n}\sum_{j=1}^{n/2} y_{2j} e^{-ijk\frac{2\pi}{n/2}} \qquad k=1,\ldots,n/2$$

$$b_k = \frac{2}{n}\sum_{j=1}^{n/2} y_{2j-1} e^{-ijk\frac{2\pi}{n/2}} \qquad k=1,\ldots,n/2$$

then

$$c_k = \frac{1}{2}(a_k + e^{ik\frac{2\pi}{n}} b_k) \qquad k=1,\ldots,n/2 \quad (7)$$

From (5) and (6) it is easy to show that a_k and b_k are periodic with period $n/2$, i.e., $a_{k+n/2}=a_k$ and $b_{k+n/2}=b_k$. In addition,

$$e^{i(k+n/2)\frac{2\pi}{n}} = -e^{ik\frac{2\pi}{n}}$$

so that

$$c_{k+n/2} = \frac{1}{2}(a_k - e^{ik\frac{2\pi}{n}} b_k) \qquad k=1,\ldots,n/2 \quad (8)$$

The splitting algorithm consists of the following two steps for computing c_k.

1. Compute a_k, b_k for $k=1,\ldots,n/2$ from (5) and (6).

2. Compute c_k for $k=1,\ldots,n/2$ using (7) and c_k for $k=n/2+1,\ldots,n$ using (8).

The splitting algorithm for computing c_k takes about half the number of operations required by (4). Step 1 requires $n(\frac{n}{2}-1)$ additions and $n(\frac{n}{2}+1)$ multiplications. Step 2 requires $\frac{3}{2}n$ multiplications, $n/2$ additions, and $n/2$ subtractions. It is customary to compare the asymptotic operation counts in which only the dominant term is retained as n gets large. Since the number of operations in step 2 is negligible compared with step 1, the asymptotic operation count for the splitting algorithm is $n^2/2$, which is half the n^2 operations required if the c_k are computed directly using (4).

The Cooley-Tukey FFT is a straightforward extension of the splitting algorithm.

The Cooley-Tukey FFT

In the splitting algorithm we note that (5) and (6) have the same form as (4), except that n is relaced by $n/2$. Therefore, if n is divisible by four, the splitting algorithm can also be used to compute a_k and b_k. That is, a_k and b_k can each be expressed in terms of two $n/4$ point transforms. If $n=2^m$, then the splitting process can be continued until the sums consist of only a single term. This process is described for the example of $n=2^3=8$ in Table 2 below.

Column 1 contains the original sequence y, which is real but otherwise chosen at random. Column 2 contains the first split into even and odd sequences y_{2j-1} and y_{2j}. The splitting process continues in columns 3 and 4. Finally, column 4 contains sequences of length 1, as well as their transforms, since by definition (4) the transform of a sequence of length 1 is just the element itself. Beginning in column 5 the transforms of shorter sequences are combined into the transforms of longer sequences using (7) and (8). For example, consider the computation of the first transform of length 4 in column 6 from the first two transforms of length 2 in column 5. In this case $n=4$, $a_1=-.62$, $a_2=1.2$, $b_1=.25$, and $b_2=.39$. Thus from (7) and (8)

$$c_1 = a_1 + e^{i\frac{\pi}{2}} b_1 = a_1 + ib_1 = -.62 + .25i$$

$$c_2 = a_2 + e^{i\pi} b_2 = a_2 - b_2 = .81 + .00i$$

$$c_3 = a_1 - e^{i\frac{\pi}{2}} b_1 = a_1 - ib_1 = -.62 - .25i$$

$$c_4 = a_2 - e^{i\pi} b_2 = a_2 - b_2 = 1.59 + .00i$$

The multiplication by 1/2 is not necessary at each step since the final sequence can be divided by n. Column 7 is then computed using (7) and (8) in which a_k and b_k are the first and second sequence, respectively, in column 6 when n=8.

Column 7 is n=8 times the discrete complex transform of column 1. In fact, each sequence in columns 5, 6, and 7 is a scaled transform of the corresponding sequences in columns 3, 2, and 1, respectively.

TABLE 2

The Cooley-Tukey FFT for N=8						
1	2	3	4	5	6	7
.07	.07	.07	.07	.25	-.62+.25i	-.05-.10i
.40	.91	.32	.32	.39	.81+.00i	-.03+.81i
.91	.32	.91	.91	.62	-.62-.25i	1.19-.42i
.18	.29	.29	.29	.20	1.59+.00i	.30+.00i
.32	.40	.40	.40	.16	.57+.16i	1.19+.42i
.56	.18	.56	.56	.96	-.03+.00i	-.03-.81i
.29	.56	.18	.18	.57	.57-.16i	-.05-.10i
.75	.75	.75	.75	.93	1.89+.00i	3.48+.00i

All computation in the Cooley-Tukey algorithm is confined to columns 5, 6, and 7, which is called the combine phase. Columns 1 through 4 are called the sort phase, and column 4 is just a reordering of column 1.

Following the sort phase, a FORTRAN program for computing columns 5, 6, and 7 would appear as:

```
      N=2**M
      DO 10 K=1,M
      NL2 = 2**K
      NL1 = N/NL2
      DO 10 I=1, NL1
      DO 10 J=1, NL2
         .
      code for equations (7) and (8)
         .
   10 CONTINUE
```

NL1 is the length of a sequence in column 5, 6, or 7 and NL2 is the number of sequences in the corresponding column. On a pipeline computer the length of a vector or sequences is determined by the length of the inner loop which is NL2. However, note that NL2 ranges from 2 to N as J ranges from N/2 to 1. Thus, computation on a pipeline computer becomes inefficient for large values of J where NL2, and consequently the vector lengths, are short. An important modification to the program given above is simply to examine the relative sizes of NL1 and NL2 and make certain that the largest value corresponds to the innermost loop. The modified code is then

```
      N=2**M
      DO 10 K=1,M
      NL2=2**K
      Nl1=N/NL2
      If(NL2.LT.NL1) GO TO 30
      DO 20 I=1,NL1
      DO 20 J=1,NL2
         .
      code for equations (7) and (8)
         .
   20 CONTINUE
      GO TO 10
   30 DO 40 J=1,NL2
      DO 40 I=1,NL1
         .
      code for equations (7) and (8)
         .
   40 CONTINUE
   10 CONTINUE
```

This modified program alone can reduce computing time by 50% on the CRAY-1. The length of the inner loop still varies in the modified program; however, the minimum vector length is now proportional to $\sqrt{n}$, rather than two for the unmodified program.

Although the modified program can significantly improve performance on a pipeline computer, the fact that the vectors still vary in length is a concern. Long vectors with the same length are possible if a large number of transforms are to be computed. If m transforms of length n can be computed at the same time, then each operation in the FFT can be applied to all transforms. The result is that all inner loops in a FORTRAN program for multiple transforms would have length m. If m is large compared with n, this approach is more satisfactory than making m separate single transforms of length n.

If m is not large, then another approach is possible. In Table 1 columns 2 through 6 describe a multiple transform problem with m=2 and n=4. Column 6 contains the transforms of the two sequences in column 2. Therefore, it is possible to compute m transforms of length n as a single transform of length mn by using a truncated version of the FFT. The minimum vector length for this approach is proportional to $\sqrt{mn}$, which is larger than $\sqrt{n}$, even for small m. This algorithm is discused in more detail in [7].

In summary, the following strategies are possible:

1. The best performance on a pipeline computer is possible when many transforms, say m, can be performed at the same time. For this case two approaches can be considered. First, each operation in the FFT can be applied to all m sequences at the same time. All vectors are of length m, so that this approach is efficient when m is comparable to or larger than n. A second approach is to transform the m sequences of length n as a single sequence of length mn using a truncated version of the FFT. The choice between these approaches is not obvious and may depend to some extent on the application.

2. The next-best performance on a pipeline computer is obtained for a single sequence that is very long. Using the algorithm suggested in this section, the minimum vector length is $\sqrt{n}$.

3. The worst case, from a megaflop point of view, is a short single transform. However, since the total computing time will be minimal, the best choice is to use existing software, which will minimize the investment of programming time.

REFERENCES

1. R. W. Hockney and C. R. Jesshope, *Parallel Computers: Architecture, Programming and Algorithm*, Adam Hilger Ltd., Bristol, England 1981 (distributed in the USA by McGraw-Hill).

2. Ronald D. Levine, Supercomputers, *Scientific American*, vol. 246, no. 1, 1982, pp. 118-135.

3. R. Reddy, Computer Science Department, Carneigie-Mellon University, From George Forsythe Memorial Lectures, Stanford University, 1983.

4. H. S. Stone, Parallel tridiagonal equation solvers, *ACM Trans. Math. Software*, vol. 1, 1975, pp. 289-307.

5. P. N. Swarztrauber, A parallel algorithm for solving general tridiagonal equations, *Math. Comp.*, vol 33, 1979, pp.185-199.

6. P. N. Swarztrauber, The solution of tridiagonal systems on the Cray-1, In *Infotech State of the Art Report: Supercomputers*, Eds. R. W. Hockney and C. R. Jesshope, Infotech Intl., Maidenhead, Berkshire SL6 16D, England, 1979, pp. 343-358.

7. P. N. Swarztrauber, Vectorizing the FFT´s, In *Parallel Computations*, Ed. Garry Rodrigue, Academic Press, New York, 1982.

KENNETH L. RATZLAFF   CHAPTER 7

Computer Science and the New Small Computer

This presentation is an essay on what a physical scientist needs to know about computers if, as is generally the case, that scientist will not be using computers as a primary research tool, but as a supporting tool for research.

This discussion must begin with a comment about why computers are used in chemistry. The computer is an incredibly versatile tool for processing input information and data, not for doing chemistry, but for aiding chemists who study chemistry. From this versatility, great variation has resulted in how these processors are used and in the types that are used. However until now, most of a chemist's training, at least at the undergraduate level, has come in the form of "CS 101", the beginning course in computer science, oriented to a narrow range of problems.

Chemists in turn deal with a very wide range of problems, the solutions to which are very often found more easily with the computer's help. How can this diversity help to define the course of study?

. The Analytical Analog

Some digression might be useful in looking at how another class of tools is presented to chemists. Instrumentation is necessary for solving most current chemical problems including such diverse questions as the structures of compounds, purity and/or concentrations in mixtures, the fundamentals of molecular structure, and the mechanisms of chemical reaction. For these we choose from a wide range of "processors": separations methods, electrochemical methods, optical spectroscopies, magnetic spectroscopies, computational techniques, mass spectroscopy, thermal methods, and many, many others.

A course in "Instrumental Methods" typically includes the following for each technique under study:
- the physical fundamentals for the particular phenomenon which is the basis for the measurement,
- the basic block diagram of a typical instrument,
- the types of problems or samples for which the technique is suited,
- the probable sources of error to be incurred, and
- some comparison with competitive techniques.

How would our Instrumental Methods course be taught if we were to attempt to teach it in the same way that "CS 101" has traditionally been taught? First we might try to find the oldest instrumental technique still in common use which can, if necessary, be used to solve a very wide array of problems, the instrumental equivalent of using FORTRAN on a centralized computer. That technique might well be titration. The student could be provided with a burette and a manual of titration methods involving acid-base, oxidation-reduction, and complexation reactions in the determination of various elements. Fundamentals of the method could be down-played since a skillful, careful titrator will get better accuracy and precision in standard determinations than someone who only knows the theory.

After such a course, we might wish to assume that if one can fully master this skill, then the discernment, creativity, and insights useful in new techniques will follow.

Few educators would be willing to take such an approach in chemistry, and that narrow approach is no more valid in educating students to properly use computers. These students must know (1) what the computer can do in the different types of applications that scientists are likely to encounter, and (2) which type of computer should be used and why.

II. The Generalized Computer Application.

The diagram in Figure I illustrates what might be called the generalized computer application. It is a diagram which may fit the application of a computer to any scientific problem if some torsion in the diagram is allowed. Centrally, there is a **problem**, a reason to use the computer in the first place. The problem requires that the computer have some **input** data and/or information. Also, the computer will generate some **output**: information which will help in reaching the research objective or data which may be further processed by the investigator, the computer, or the instrument.

Computers to be used in the solution of this problem have all of the variety of the other tools used by scientists. A brief overview of the richness of the area is useful in identifying topics for study in an appropriate computer-related course.

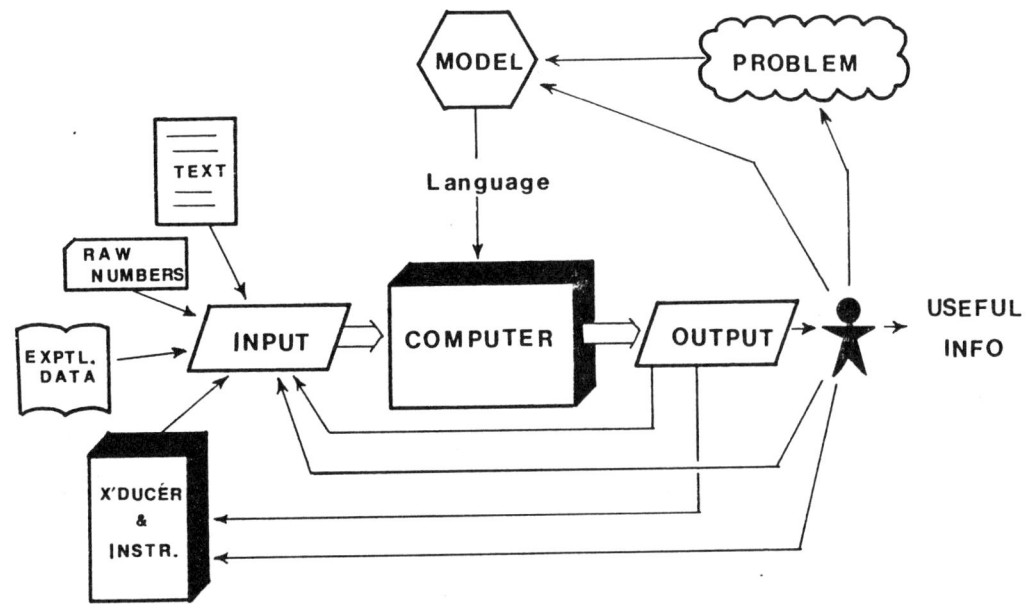

Figure I
The Generalized Laboratory Computer Application

II.A.   Problem Types.   Three categories are sufficient to handle most computer-oriented problems. They are differentiated by the types of input and output variables.

The first category, for which computers have been used the longest, are those which are Computation Intensive. Examples include many types of modelling, including ab initio calculations and studies of models of the atmosphere or instrumental characteristics. There are several properties which greatly influence the approach: some computations are amenable to formulation in matrix form and some are not; some are completed exactly, and some require iterations.

Some equally large calculations can be called Data Intensive. Large data bases must be manipulated to find certain items or to make correlations. For example, scientists are making increasing use of on-line literature search facilities in which the user is very seldom involved in any level of programming the computer. Inventory control would also fit into this category.

Statistical methods such as factor analysis are often used by environmental scientists and some biologists. Typically, the the investigator writes the program to format data but uses proven subroutine libraries.

A third type of problem involves **Real-time Experimentation**, the use of the computer on-line with an experiment. The data are acquired from the experiment, and the output is often fed back to the experiment. At one level, all of the now-common microcomputer-controlled, "intelligent" instrumentation falls into this category, but the programs are all "canned", resident in the instrument and unavailable to the investigator for examination or refinement. The investigator relates to that instrument in the same way as to older instruments. The real-time use involves taking immediate advantage of the computer's versatility in order to become involved in a variety of experimental input and output values.

II.B. Computer Inputs. The types of inputs are as varied as the problems. Most applications involve the processing of discrete items of data which can be divided into three categories. In other cases the input form is less obvious.

The first category of input is **Transcripted Data**. The data, having been collected off-line, are later entered into the computer at a keyboard or, in the case of graphical data, a digitizer. These data sets can be experimental data derived from instruments or direct observations; alternatively, they might be text when the computer acts as a word processor. In the latter cases, the data sets may be large, e.g., from the investigator's archives, but often they are relatively small. In some other cases, the data is retrieved from a large data base, either shared with collaborators or obtained elsewhere.

A second data source is **on-line instrumentation**. With most newer intelligent instruments, the instruments are themselves responsible for logging the data and doing preliminary computations, and the small computer, treating the instrument as a "front-end input device", then acquires that data through a communication process. Although chromatographs and spectrometers come first to mind, it is not uncommon for a microcomputer with graphics capability to send input to and retrieve output from, for example, a Cray supercomputer, treating it as simply another type of instrument.

The third case brings the small computer closest to the experiment. Through real-time peripherals, analog-to-digital converters, digital-to-analog converters, clocks, and digital interfaces, the small computer becomes the basis of the instrument, dealing directly with the **Transducers** which move data between the electronic and the chemical/physical domains.

Frequently there are few of what are conventionally considered inputs, particularly in modelling computations; the inputs are the model itself, often in the form of a program with few if any "READ" or "INPUT" statements.

II.C. Computer Outputs. The outputs from a computer may be either data for distribution or further processing, or they can be answers. Several categories can be considered.

We are most familiar with sorted and processed data. Grade sheets from a university or test results from comprehensive testing in an analytical facility might be examples. At other times quantitative answers are required, for example, "Yes, with 99% certainty we can say that benzene is present". Descriptive information can sometimes be provided. A factor-analyzed set of complex data can lead to direct conclusions.

Graphical output is very important to many scientists. A drawing of a crystal structure or a plot of a spectrum conveys far more information in a short period of time than does any table of numbers.

Much of the output from a computer must be classified as intermediate results. In a long-term experiment, the data serve to aid the investigator in the design of the next experiment while in the short-term on-line experiment, data is immediately fed back to the experiment in the process of control, optimizing experimental variables to improve the quality of the data.

II.D. **Required Response and Execution Times.** The response requirements define the maximum time that can be allowed for the computer to respond to some input. They fall into about four categories. When **Batch** response is sufficient, the inputs will not change before the response is obtained; the program and inputs can be submitted to the computer, and the results can be obtained the next day. Somewhat faster response is required if **interactive** processing is desired; in this case, the computer program responds to information entered at the terminal during program executions, and at the very worst, a few seconds can be tolerated after a line is entered before the computer makes some response lest the operator lose patience, although no data need be lost. Slow **real-time** response is sufficient if the period between the availability of data points is on the order of seconds or longer so that the computer need not devote its entire attention to the input; even so, the data will be lost if the data is not acquired in the allotted time-frame. Fast real-time response is required for data acquired in the kilohertz and greater time frame.

The execution/turn-around time is a slightly different problem. Batch programs can be short or long. Other types of applications may simply require the computer only to monitor the input and store the data, an application which might take only a small fraction of the computer's attention; other input devices require continuous control.

III. **The Available Tools.**

At the basis of making the optimum use of the computer is the choice of the right tool for the application. What are these tools?

III.A. **The Hardware.** Computers are categorized roughly by size which _may_ be related to power. Most research facilities and educational institutions have a **central computing** facility, a computer which is large, remote, and shared making it nearly inaccessable on a real-time basis. Within a department there may be a large mini- or "midi"-computer. This computer is more likely to be used flexibly; it is quite possible that one will be able to negotiate with the operator for slow real-time response on a time-shared basis. The smaller **mini-computers and micro-computers** (in operation and concept there is no difference), have a very small number of users, usually one; the investigator may have complete control of how it is used. The microcomputer configured without operating systems or a variety of high-level languages is usually a **micro-controller** with fast response but limited flexibility and power.

At several of the above levels, the **Central Processing Unit** (CPU) can take a variety of forms. We all know that some CPUs are 8-bit models, others 16, 18, 32, 64, 72, etc. In actual fact, the distinction is not so clear; most 8-bit CPU's execute some 16-bit instructions, and 16-bit CPUs have 8-bit and 32-bit instructions. Some computers have only one CPU while others have several. Among the latter, the CPUs may operate in parallel, for example, by using separate math and I/O processors, or as is the case with the Apple/softcard combination, the CPUs simply substitute, the appropriate one being chosen for the job. Vector or array processors can be used to great effectiveness in certain applications.

Some computers have separate fast and slow **memory**, and cache memory and virtual memory architectures are often useful. However, possibly the key factor concerning memory is that it is rapidly becoming so inexpensive. There has always been a tradeoff between writing compact code on one hand and saving development time on the other. This tradeoff is becoming heavily weighted in favor of the latter.

Peripherals cover a wide range. For mass storage, everything from cassette tapes to hard disks is used. For user I/O, we have CRT terminals, dot matrix and word-processing printers. Graphical output is the most immediate form of user output. Serial and parallel interfaces communicate with instruments. Laboratory peripherals are used to deal with transducers.

III.B. Software. Software can be divided between the application software, the development tools, and the operating system. The application software is that which the user actually executes, the program which calculates the least-squares fit or controls the instrument. The development tool is usully the language in which the application program is written. The operating system carries out the routine house-keeping necessary for the convenient use of the computer.

Some application software, for example, word-processing software, is used directly; the package is purchased and executed by operators who do not prepare any of the program. Most scientific applications require a program to be written locally, and that requires that the appropriate development tool be chosen.

The languages which are available on most computers each find their niche in certain applications. FORTRAN is historically the primary scientific language; its usefulness continues because of its standardization and the huge software base left by other programmers. BASIC is available on nearly every computer now available; application programs, though slow in execution, are easy to write and, for short programs, quick to debug. Pascal's virtue is its orderliness; the code is easily debugged and easily maintained. C is a high-level language which has low-level functions making it useful for such problems as writing word-processors, communications software and operating systems; it is sufficiently standard to make the software highly portable. FORTH is an extremely powerful language for real-time experimental control; once fundamental definitions are set, programs are developed and executed very rapidly. Among the many lesser languages in some scientific use there are Ada, LISP, APL, and ALGOL. PL/I, designed to be all things to all people (the programmer's Swiss army knife), could possibly serve a wide range of applications. Operating Systems may be designed for batch processing, multiple-users, or single users, single or multi-tasking, and they may be optimized for user interactive operations or real-time input/output. The multi-user/time-sharing approach approach which was the most cost-effective in the past is giving way to networking software; each user has a computer, and only certain resources (data-bases, printers, plotters, mass storage, etc.) are shared.

IV. An Example.

Many of the problem types and tools are brought together in an application involving a rapid-scan spectrophotometer. In this instrument, a solid-state detector array placed on the focal plane of a monochromator (Figure II) allows intensity data at 1700 wavelengths to be obtained simultaneously, making a 250 nm spectrum available every 30 ms. In the data acquisition stage, the computer must control the operating parameters and obtain data at about 100 KHz. The data can be tested immediately to determine the necessary changes in operating characteristics. The raw absorbance data forms a matrix which, if an array processor were available, could be rapidly processed. When the processing is complete, graphical output is mandatory if the data are to be interpreted in a reasonable period of time.

After the data are gathered, they can be subjected to a variety of processing techniques. For example, the data set might be massaged by use of an automated, weighted least-squares fit in matrix form allowing very low absorbance spectra to be extracted from various sources of noise. Further computations involving Fourier-transform correlations can be used to search data bases for the same or similar spectra.

In a single laboratory application, the computer must act as a fast controller, must search data bases, and must perform a host of computations.

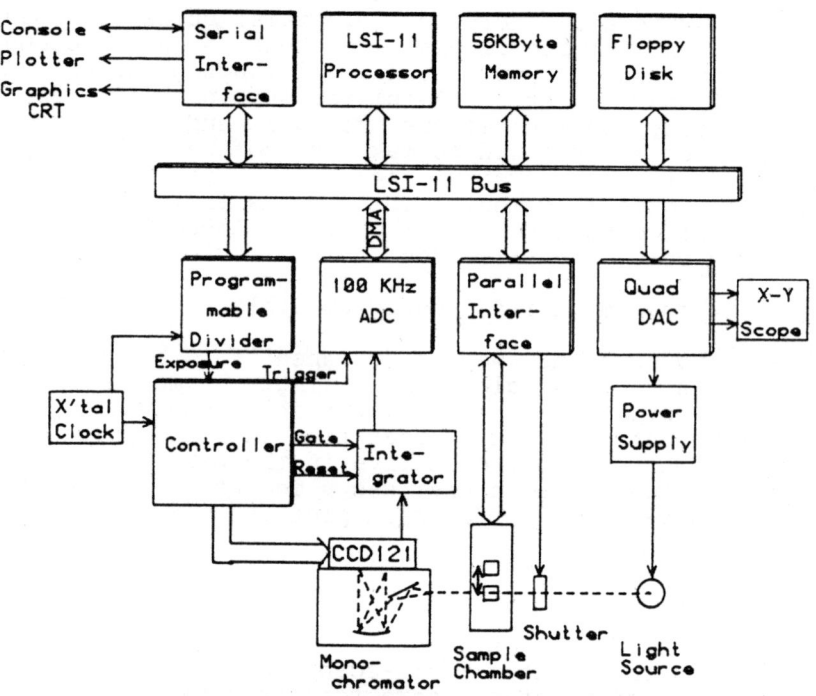

Figure II
Block Diagram of a Rapid-Scan Spectrophotometer
Based on an Optical Imaging Device

V. What Must Chemists Know About Computing?

This essay has stressed the tremendous variety of scientific computer applications. It has also stressed the fact that the small computer, often microprocessor-based, can and will soon be asked to handle most of those applications, either alone or in networks. It is this computer to which today's students will relate most often.

Clearly it is impossible to prepare a student to immediately carry out all possible applications. Given that most students will use computers in their careers but not make a career of using a computer, what questions should they be able to answer after an undergraduate education?

First, what kinds of problems are amenable to assistance from the computer? In most scientific environments, knowing what type of help is available solves half the problem because then the investigator knows with whom to collaborate or where to get help.

Second, what computer system characteristics are necessary to solve that problem? Are there hardware/software tradeoffs? What software tool should be used?

Third, which computers have the necessary characteristics? Does one request computer time on the central mainframe computer or request an allocation for a microcomputer? It is very interesting to note that nearly all of the application classes can be handled by some type of "microcomputer". There is little doubt that within two years, computers based on <u>currently-available</u> microcomputer boards (CPU's, memory, and array processors) at a hardware cost under $20K will outperform midi-computers such as the VAX 11/780 in many applications.

Fourth, how does one program in a disciplined and efficient manner? In many cases this question suggests that the language to be taught at the undergraduate level is <u>Pascal</u>; former <u>Pascal</u> programmers probably become better <u>FORTRAN</u> and <u>BASIC</u> programmers afterwords than they might otherwise become.

Fifth, what are ways of avoiding non-programming errors: aliasing of data, incorrect filtering, numerical instability, improperly applied algorithms.

Finally, although it may be necessary to learn it, computer jargon should be avoided. The mystique around computers must give way to a relaxed relationship for all users.

An appropriate course will stress these basics, avoiding the details of computer techniques in which they will become only secondarily involved. For example, the algorithm used in a program for performing a literature search on a remote computer where the investigator only operates as a user has little importance. The means of programming and interfacing a small computer to communicate with an instrument, a task which a scientist is likely to face, is fundamental.

Topics which might make up such a course include the following:
1. Modes and basic definitions in computing.
2. Interactive programming.
3. Basics of computer architectures.
4. Real-time computing using standard interfaces.
5. Input devices
6. Languages, a survey.
7. Graphics.
8. Mathematical techniques in computer programs.

In the real world of the physical sciences, the use of computers involves much more than the development of programming skills, and ways will have to be found to reflect that difference. Unfortunately, Computer Science departments are seldom oriented to those needs; chemists have taken the lead among scientists in this area and need to continue to do so.

HOWARD SALTSBURG, RICHARD H. HEIST, AND THOR OLSEN

CHAPTER 8

The Microcomputer in
the Undergraduate Laboratory

INTRODUCTION

There is little disagreement that the computer is a tool of immense value to the scientific community, but there is significant variation in the manner in which this tool is introduced to a beginner, whether a student, in the course of study, or a working professional, catching up. Although numerical computation is the prevailing use, it is becoming increasingly difficult to find instrumentation which is not interfaced to the computer. In the research laboratory, significant effort has been devoted to interfacing of hardware and development of software for data logging and analysis. On a larger scale, chemical plants are being placed increasingly under digital control. The cost of these systems, both in time and money, is not insignificant.
In addition to the obvious uses outlined above, programming itself is a form of problem solving. Techniques learned in the study of programming concepts (not syntax), such as problem formulation and decomposition, are transferrable. The structure of the many available high level languages provides a variety of logical approaches to problem solving.
The problem of introducing a student population to the use of the computer in a meaningful way is of

concern to most educational institutions. The traditional numerical analysis approach is adopted without too much difficulty as most professionals show little concern about the time invested: the return is clear. The alternative approach -- through the laboratory -- is usually not considered educationally viable. The relatively dedicated nature of laboratory computers makes the purchase of equipment (including the interfacing devices) an apparently less effective use of resources. The lack of appropriate software appears to be a formidable obstacle, and it is clear that significant effort is involved, particularly if the interfacing hardware must be designed and constructed by the users. Finally, although the cost of using a remote facility via a terminal is often minimal for instructional purposes, laboratory costs including computers tend to be local and born by the immediate users. Professionals, initiating laboratory computing, face similar problems.

For the past four years the Department of Chemical Engineering at the University of Rochester has been involved in a program to introduce the computer into the curriculum via the laboratory. In the following, the rationale for the approach which was adopted will be presented together with some selected examples taken from the laboratory program. The case for the microcomputer in this role will be developed (1).

THE APPROACH

If one considers using a computer simply as a data logging device, for example, for measurement of temperature or the bridge output signal of a gas chromatograph, several properties of the desired data must be decided upon. The most immediate are the precision of the data that is needed, the speed of data acquisition required, and how much data must be stored for eventual analysis.

Using the chart recorder as a measure, a precision of no more than 0.1% of full scale is possible. This is represented by a digital precision of 10 bits; 0.25% can be represented by 8 bits. For most laboratory applications, the required speed of data acquisition is low by modern electronic standards. Further analysis indicates that an extensive dynamic range often is not needed, and when it is required, slow range switching is appropriate. Data storage can be a problem, but the slow nature of the signal acquisition often permits peripheral storage to be used effectively.

Given that most needs are for slow, low precision data handling, how can one use this fact to "get into the game"? What is needed is sufficient understanding at a level sufficient to make use of the computer as a tool. There is no need to become a designer, and it would be inappropriate for most users. Fortunately, there is a literature directed at the builder and user. The focus of this part of the amateur or hobby literature, a successor to the radio amateur literature, is on simple and inexpensive solutions to hardware problems in which the demands for precision and speed are modest; more particularly, those situations where 8 bits and response in tens of milliseconds will suffice. The user oriented discussion of device operation is written at a level which does not require that the reader have a sophisticated electronics backround. Logic-level based discussions often will be adequate as the integrated circuits (which make all of this possible) are also described in this manner. In addition, there are software discussions, also at this user level, which include details of programming practice, programming aids, and descriptions of operating systems. Discussions of complex topics at an elementary level designed for the beginner, as well as more advanced material, can be found in these books and magazines. Although hobbyists do not count the time expended to achieve a task, professionals to whom the computer is a tool may find that such expenditure of time would be inappropriate. The sharing of this information via these publications thus makes an important resource available to a wider audience (2).

HARDWARE CHOICES

The traditional choice in the laboratory has been the terminal connected to a mainframe computer. Analysis of the typical user requirement shows, however, that there is little need for the capability of the mainframe in an undergraduate laboratory. The need for the user to become involved with a complex operating system before anything can be done, the fact that system down-time is not under local control, and the operating expense, all were factors which combined to rule out this approach. Real-time needs are not best served by a time sharing operation, and even the use of a local minicomputer is not an optimal solution.

At the other end of the hardware spectrum are the computers known as single (bare) boards. These are exemplified by the KIM, SYM, and AIM (3). Significant time is required to build up the system to the point where a beginner can use it effectively. It is difficult to achieve the quick return on expended

The Microcomputer in the Undergraduate Laboratory 157

effort which is required to keep up the interest while learning. Moreover, most of these units do not have the sophisticated file handling firmware or software enabling convenient use of peripheral I/O devices for data and program storage. High level languages are an additional component since machine coding is the common programming mode for these units.

Self-contained microcomputers appear to be nearly ideally suited to the laboratory task. They are equipped with a high level language, typically BASIC, video display and keyboard, and they have readily accessible data and address busses, and often sophisticated I/O facilities. The latter make interfacing much simpler and safer. Machine language monitors are readily available or are part of the operating system. File handling is incorporated in the operating system, and inexpensive peripheral program and data storage is readily available. Also, most micros can be used in terminal mode, e.g., when data need to be transferred to a more powerful system for extensive computations.

The cost of the self-contained microcomputer is so low that multiple units are feasible and a computer can be dedicated to an experiment. The total cost of a comparably equipped single board is actually higher, and the board often is not as well supported in terms of the available user-oriented literature. Perhaps the most important pedagogical advantage of these devices is that the system is personal and highly interactive so that reinforcement is immediately available; when all else fails, the power can be turned off and a fresh start made.

Real-time graphics in high resolution is not generally available on microcomputers, but this is not a significant problem. The need is for sufficiently fast response so that the user can make appropriate decisions as to the subsequent course of events. This time --"useful time"-- is a more important parameter. If one can wait a few minutes, video graphics or high resolution graphics hard copy can be generated in useful time. The dot matrix printer operated in the graphics mode becomes an important part of the system.

INTERFACING

The formidable task of interfacing is a common deterrent to the introduction of computers into a research laboratory. As noted above, however, the requirements are not often as severe as assumed, particularly in a teaching situation, and simple interfacing devices of the type found in the amateur literature will perform adequately for many purposes both in research and instructional applications.

An example of the utility of the amateur literature in this regard is the description of a simple joystick interface to be used in cursor manipulation with games (4). Since the joysticks considered were potentiometer-based, the circuit will also serve to interface other resistance-based sensors. The essential component of the interface is an integrated circuit timer which can be operated as a triggered pulse generator, the length of the pulse being determined by the resistance and capacitance of the external circuit: a fixed capacitance makes it a resistance digitizer. Appropriate software permits one to use the internal timers of the computer to measure the pulse length, thus reducing the cost and complexity of the interface. One chip, two capacitors and the resistor (sensor) are all that is needed. Calibration of the sensor completes the interfacing operation.

This interface, used with a thermistor, provides a very inexpensive and easy-to-understand device for the measurement of temperature. A precision of up to 10 bits can be obtained. What has been given up is speed of response. (In fact, the trade of speed for precision marks many of the inexpensive A/D converters.) It should be noted that the characteristics of the timer are well documented in the electronics literature, but the amateur literature makes this information available in a particularly user-oriented form. In addition to the thermistor, a photoresistor can be interfaced using the same circuitry and software. Such a sensor has become the basis for an inexpensive but extremely useful colorimeter (5).

Voltage based A/D converters are needed when sensors such as thermocouples must be used. Simple devices of this type have also been described in the amateur literature, the simplest being pulse-width based (6). The latter use the same software for data acquisition as the timer IC based resistance converter, simplifying their integration into the laboratory. Voltage converters require a higher signal level than most sensors produce so amplification is required. This is an additional complication, but the amateur literature again describes useful, simple circuits (7).

THE CHEMICAL ENGINEERING LABORATORY PROGRAM

To introduce the computer into the laboratory program, it was clear that one should start with a computer course. It seemed unwise to place students in

The Microcomputer in the Undergraduate Laboratory   159

an environment of computer-aided experiments without some preparation. The introductory course was to cover the basic background and provide an opportunity for both faculty and students to learn. It also permitted a slow and orderly transition into the changing laboratory program. Since our laboratory program is composed of nominally independent courses, it was possible to create a new sophmore computer course. The consequence of this approach is that it took three years for the senior laboratory to contain a significant number of computer interfaced experiments. The ground rule for the inclusion of the microcomputer in a given experiment is that the experiment must be improved. That improvement may take several forms. It may make a new parameter regime accessible, it may reduce tedium of data acquisition or reduction, or it may permit new experiments to be developed. Examples of all are found in the University of Rochester program.

The sophmore course covers the elements of computers, computing, and interfacing. It is general enough to serve a wider technical audience. The computer is described in terms of what it does, how it works from a logical system point of view and how these functions are implemented electrically. Structured programming is used in the form provided by Waterloo BASIC for the Commodore PET/CBM (3) machines. Machine coding is introduced as required to deal with the data acquisition devices. Elements of I/O operations are illustrated by the turning on and off of light emitting diodes. This simple example of on-off computer control provides strong reinforcement of the principles and there is a clear recognition of success and failure in carrying out the assigned task. Analog transducers are described along with A/D and D/A conversion techniques. The final project is to write a structured BASIC program to control the temperature of a recirculating air heater. This device is sufficiently inexpensive to permit the use of one device for each available computer, so each student can work alone, and it is sophisticated enough to provide meaningful control options (8). It should be noted that closed loop process control is normally given to senior students, and rarely has a significant laboratory component. In the sophmore course, only a cursory introduction to closed loop control is given in the form of the appropriate algorithm for proportional control. It has been found that students not only were able to carry out the assigned programming task using proportional control but that some were able to discover other, more sophisticated, algorithms (PI,PD) within a few hours. The rapid discovery is a result of the fact that the device responds quickly enough so that exploration is possible without an excessive wait. One can learn by doing and do it quickly. This is what laboratory studies should accomplish if they are to progress from simple exercises to a true learning environment. The ability to have a reliable working model which has the elements of a realistic problem is a consequence of the computer aided experiment. Interestingly enough, the senior students studying process control are now able to be given programming assignments using the air heater, and they have shown that much more sophisticated algorithms can be implemented and tested (11).

Students do not build any hardware nor are they required to do extensive machine coding. They are expected to understand the hardware and software operation as users. It is our intention to develop enough understanding so that our graduates can communicate intelligently with design professionals. That is not possible in one introductory course, but as more use is made of the computers in the subsequent courses, better understanding is developed.

The junior laboratory program deals with principles of interest to chemical engineering, including chemical kinetics, thermodynamics, heat transfer, and fluid mechanics. To date there has been no attempt made to interface the fluid mechanics experiments to the computer as there was little advantage to be gained with the systems in use. All the other areas have been affected by the computer.

One very important application has been the interfacing of the gas chromatograph to the microcomputer (9). The reduction in tedium is sufficiently important that the computer now is regarded as part of the GC system. Student response to this aspect has been (predictably) good since the GC is an essential analytical tool in the program. It is not uncommon for twenty or thirty chromatograms to be produced during the course of a single laboratory session and there is no loss in content in having the area under the peak presented as part of the output. The student still must deal with the reduced data: the program does not carry out the reduction to the final required analysis.

Another area in which the computer has had a significant impact is the measurement of temperature, particularly in transient studies. A common measurement is of the thermal diffusivity of solid materials. The procedure involves following the temperature at a point in an initially isothermal solid when a step change in the boundary temperature is imposed. For analytical simplicity, a sphere with a

The Microcomputer in the Undergraduate Laboratory   161

temperature sensor at the center is often chosen. In the past, using slow data loggers (e.g. a chart recorder), in order to obtain adequate resolution it was common practice to use a large sphere to slow down the entire process of equilibration. This limited the extent of the study. Using the computer, the complete study is now carried out in about half the time and with five materials including an aluminum sphere. The sphere is taken from an ice bath and placed in boiling water: the aluminum equilibrates within a few seconds but the computer easily records one hundred data points. The transfer time between the two baths has a measureable effect on the initial temperature of the sphere when it enters the second bath. This experiment then permits one to explore the heat transfer process to the sphere as well as the thermal respone time of sensors.

The senior laboratory is concerned with the integration and application of these concepts in terms of process devices. Thus, separation processes, such as distillation, and chemical reactor operation and performance are investigated. The systems whose properties were studied in the junior laboratory are used whenever possible in the process devices. The chemical reaction studied in the junior laboratory, where the objective was to learn about the kinetics of the reaction, is used in the flow reactor study, where the objective is to learn how the configuration of the reactor influences the overall conversion achieved.

Studies of the operation of a distillation column require that the temperature of the trays as well as the boiler be monitored. The ability to measure temperature using thermistors made it possible to provide an inexpensive, multichannel digital themometer for the device. As a result of the continuous display of these temperatures, which are updated every few seconds, and the recording of the changes which have occurred between successive readings, column dynamics are much more readily followed. Students are able to see the changes in the tray temperatures as the operating parameters are deliberately varied (e.g. boiler heater) as well as the effect of suddenly cooling the column walls. One of the questions which must be answered is when the column may be considered to be at steady state. The students noticed when the tray temperatures became constant and that is now used as the signal for tray composition to be measured. The previous technique was to sample the tray contents and analyse them with a gas chromatograph (without the computer) until they were steady. The overall improvement in the study is significant.

Once the process devices are interfaced, process

control can be added and control studies become an extension to the device study. Such control now has been implemented on the distillation column (10,11). More complex reactors have also been interfaced using the same techniques and are now under computer control.

THE LED COLORIMETER

An interesting example of the utility of this approach to the laboratory is the development of a simple colorimeter based upon the ease of interfacing of the photoresistor as an optical sensor. The device consists of a current controlled light-emitting diode (LED) and a photoresistor. The interface uses the timer IC circuit and, of course, the same data acquisition software as is used with the thermistor. The first experiment in which this colorimeter/interface combination was used was a study of the reaction between phenolphthalein and excess base. Under these conditions the pink quinoid is rapidly formed, but this reaction is followed by a slow reaction of the quinoid with hydroxide ion to yield the colorless carbinol form (12). It is a simple matter to study the reaction colorimetrically as a function of concentration and temperature. The time resolution is limited by the converter used here, but even when the total reaction time is measured in minutes, sufficient data can be acquired. Since that data can be stored in machine readable form (on a cassette tape in our case), even though the computer makes much more data available, data reduction and analysis are considerably simplified. The student, therefore, can spend time on understanding scientific and engineering principles rather than on the drudgery of data reduction.

The success of this colorimeter system led to an immediate extension to an equilibrium study of the nitrogen dioxide/nitrogen tetroxide system. The colorimeter was essentially the same, the interface was the same, and the data acquisition software was the same. Only the application program had to be changed.

The size of the components used in the colorimeter is such that it is possible to build them into a collar which can be placed around a translucent tube. This colorimeter is ideally suited for those flow studies in which a non-flow-intrusive sensor is desired, and color can be exploited. For example, a continuous flow, stirred tank reactor using the phenolphthalein fading reaction was interfaced using this colorimeter. It is an enormously simplified setup compared to the earlier device which used complex colorimeters and required much more care to obtain any useful results. An ion exchange experiment was altered to use a colored ion, greatly simplifying the study. Dispersion of a dye pulse in a very long tube has been made a possible laboratory study by using the LED colorimeter. Eight of these collar sensors installed around a translucent tube have permitted, for the first time, a simple experiment which demonstrates, with almost textbook precision, the dispersion curves. Again the analysis is simplified by the availability of machine readable data. In no case was the basic principle involved in the study compromised. The use of a colored ion in the ion exchange experiment was a case in which the experiment was matched to the computer system.

In the examples in which time dependent data are desired, graphical output on a dot matrix printer is available. This provides the high resolution display in useful time. The output can range from raw to fully processed data. It is important to note that once the program for the printer was developed, extension to new data representations was relatively simple. Thus, the program which is used to print the gas chromatogram was readily extended to print spectra taken with a UV-VIS spectrophotometer. The utility of the computer was further demonstrated by making the single beam instrument function as a (pseudo) double beam device. The lamp spectrum was prerecorded and when the sample spectrum had been recorded, the computer was used to calculate the absorbance which was then printed in the conventional representation. Spectrum manipulation (e.g. expansion, subtraction) also is possible since the absorbance data are available either in the computer or in machine readable form on tape.

OTHER CONSIDERATIONS

The high level language which is common to most microcomputers is BASIC. An unfortunate characteristic of this language is that it tends to yield code with a logical structure, or flow, which is difficult to follow. For this reason, many educators prefer the use of Pascal which is tightly structured to force good programming practices. However, Pascal is usually disk based, and the cost of disk drives can exceed the cost of the computer. To reduce cost, a tape cassette is used in our laboratory program, and this choice restricts the languages which can be used. Waterloo structured BASIC for the PET/CBM computers was chosen because it is available as a plug-in ROM and provides a convenient transition to the pedagogically preferable structured languages while retaining the user friendliness of BASIC.

An attempt to use FORTH was made, but the lack of a ROM based system made the use of this language inconvenient, and it is now reserved for advanced, special projects. The appearance of ROM based FORTH systems for the VIC-20 and the Commodore 64 (3) may alter this situation in the future.

Although the program began in the undergraduate laboratory, the demonstration of the ease with which limited laboratory operations can be handled has led to the incorporation of the microcomputer into research applications both within and external to the department. The ready availability of the computers has produced more interest in typical business use such as word processing, Visi Calc style budgeting, and automated handling of mailing lists. Monitor-equipped classrooms are now available as more instructional use is being made of the computer in the nonlaboratory courses. The fallout has been significant.

CONCLUDING REMARKS

The incorporation of computers into the laboratory is feasible and desirable. While it is of some importance as to which computer is selected, it is more important that a selection be made and the interfacing process begun. The personal aspect is critically important: one should have access at nearly any time and be able to learn at one's own pace. The interactive aspect is also vital to staying with the task of learning to use the computer. Video games provide ample evidence of that feature.

The amateur literature is an important resource which has been ignored by many educators and professional industrial scientists. Since it is written for users and builders, it permits one to minimize the start-up time in bringing the computer into the laboratory.

The example of the same interface being used for both optical and thermal sensors, with the advantage of the same software for data acquisition, illustrates the great utility of the computer as a "virtual" instrument when properly used. In effect, the interfaced computer becomes whatever instrument the software configures it to be, and thus the interfaced computer can be considered to be a universal device. The use of standardized interfaces make extensions much simpler. Little customization is needed to extend the use to another experiment. Only the applications program need be changed to reflect the physical or chemical phenomena being measured. Maintenance expenses are also reduced.

The lack of a real-time, high resolution video display is not serious under the conditions of most instructional and many research laboratories. What is needed is display in useful time (i.e. soon enough to know whether the experiment needs to be repeated). This information can be generated within a few minutes. The resolution question can also be addressed in terms of useful time. Since hard copy is often desired, a dot matrix printer can be used to provide high resolution graphics. The most elementary system consisting of computer, printer, tape cassette, and interfaces, then functions minimally as a smart chart recorder, but at considerably lower cost.

Careful examination of the goals of a study often permits one to alter the experiment to suit the equipment without loss of content, and a program can often be improved using the computer to make new experiments feasible. The available features include fast data acquisition, reduction of tedious data handling, and highly simplified instruments such as the LED colorimeter. Often, replacement of a thermocouple by a (more easily interfaced) thermistor requires no significant sacrifice, and the computer can handle the nonlinearity of the resistance-temperature curve. The curve fitting is not different, in principle, from the use of thermocouple tables.

As the cost per unit drops and the capability is increased, the microcomputer will have a significant impact on the efficiency of a laboratory program. The fact that computer interfaced experiments work and can be understood, coupled with increased productive output, can make the laboratory a very important component of the educational experience.

BIBLIOGRAPHY

1a. "Microcomputers in a College Teaching Laboratory" Part I, H. Saltsburg, R. H. Heist, T. Olsen, MICRO $\underline{53}$ 53 (1982)

b. "A Microcomputer-Aided Chemical Engineering Laboratory" H. Saltsburg, R. H. Heist, and T. Olsen in Proceedings of the Engineering Foundation Conference "The Undergraduate Engineering Laboratory" (in press, 1983)

2. Typical examples include MICRO, Microcomputing (Kilobaud), COMPUTE!, Computers and Electronics (Popular Electronics), and books in the Blacksburg Press series and the Howard Sams series.

3. SYM and AIM are registered tradenames of Synertek Systems Corporation and Rockwell International, respectively; KIM, PET, CBM, VIC-20, and Commodore 64

are tradenames of Commodore Business Machines, Inc.
4. "Hello World" J. Sherbourne, MICRO <u>26</u> 31 (1980)
5. QM-100. Analog Systems, P.O. Box 35879, Tucson, AZ
6. "Microcomputers in a College Teaching Laboratory" Part II, R. H. Heist, T. Olsen, H. Saltsburg, MICRO <u>55</u> 59 (1982)
7. "IC OP-Amp Cookbook" W. G. Jung. Howard W. Sams, Indianapolis, IN (1979)
8. "The Microcomputer in the Chemical Engineering Laboratory" R. H. Heist, H. Saltsburg, T. Olsen, Computer Applic. In The Laboratory <u>1</u> 28 (1983)
9. "Microcomputers in a College Teaching Laboratory" D. Graves, R. H. Heist, T. Olsen, H. Saltsburg, MICRO <u>57</u> 89 (1983)
10. "Microcomputers in a College Teaching Laboratory" Part III, T. Olsen, H. Saltsburg, R. H. Heist, MICRO <u>56</u> 38 (1983)
11. "A Model Program for Undergraduate Education in Real Time Process Control," T. Olsen, R. H. Heist, H. Saltsburg, J. Friedly, IFAC Symposium on Real-Time Digital Control Applic. (in press) 1983
12. "Alkaline Fading of Organic Dyes" R. P. Andres and L. R. Hile Chem Engr. Education pg.18 (Winter 1976)

J. W. SCHILLING   CHAPTER 9

Teaching Computers to Chemistry Students in a Liberal Arts University

INTRODUCTION

In 1973 the National Science Foundation awarded to Trinity University a grant for the purchase of a minicomputer so that a course called "Computers in Chemistry" could be established. At that time the microprocessor revolution had not yet occurred and it was not possible to purchase a standard "scientific" minicomputer system for under $20,000; I opted to purchase a Datapoint 2200, a business computer, and to build the equipment required for an interfacing laboratory.

The goal of the proposed course was and is to provide a degree of "computer literacy" appropriate for baccalaureate students in chemistry. A bare minimum requirement for this would be some knowledge of BASIC or FORTRAN, preferably both. However since chemists work at both the hardware and software levels, computer literacy should include more than just a knowledge of programming. Students should also understand how computers work electronically in order to be able to interface them to laboratory instruments, and how they work logically, i.e., at the assembler language level. The latter is considered by many to be unnecessary today, but I feel it to be important so that students will understand the limitations of higher level languages, and so that they will be able to make intelligent choices when it is necessary to specify the purchase of equipment. A chemist can not (and should not) try to do the job of an engineer, but he or she should be able to talk intelligently to one.

THE INSTITUTIONAL AND DEPARTMENTAL CONTEXT

The strategy adopted at Trinity grew out of our own institutional and departmental environment, and it is difficult to evaluate the work reported here without an overview of that environment. Trinity is a liberal arts university committed to primarily undergraduate instruction and a small number of selected graduate programs. There are about 2600 undergraduate and 650 graduate students. The department has six full time members, expanding to seven in 1984. The enrollment in General Chemistry is about 150. An average of two B.S. majors and four to five B.A. majors graduate each year. Of our B.S. majors, approximately 75% go to graduate school and the rest go to business and industry. The Chemistry Department until this year has offered a traditional research oriented Master's degree; this program has attracted virtually no students for some years now, and we have revised it to focus on chemical instrumentation.

The liberal arts ideal precludes the degree of specialization at the undergraduate level that might be possible at a technical university. Table 1 shows the chemistry curriculum. The B.S. degree is approved by the American Chemical Society; the B.A. degree would be chosen by pre-med students and those wishing a double major in chemistry and some other field.

INTEGRATING COMPUTER SCIENCE INTO THE CURRICULUM

The major stimulus of the work reported here is the difficulty of introducing computer experience into the B.S. program. Something must be deleted for every new requirement. Because of this curricular pressure chemistry majors at Trinity are not required to take a computer course; neither are they required to take courses in other desirable areas, such as statistics, linear algebra or electronics. These must all compete for a very limited number of elective hours with other worthy subjects. Nevertheless we support the liberal arts concept; it is not an obstacle to be overcome but rather an important competing priority.

TABLE 1
B.S. and B.A. Chemistry Curriculum at Trinity University

B.S. PROGRAM	B.A. PROGRAM
General Chemistry	General Chemistry
Organic Chemistry	Organic Chemistry
Physical Chemistry	Physical Chemistry
Advanced Chemical Techniques	
Molecular Spectroscopy	
Instrumental Methods	
Advanced Inorganic	
One other advanced course	Two advanced courses
Three semesters of math	Two semesters of math
General Physics	General Physics
Russian or German	

B.S.		B.A.
40 hours	CHEMISTRY	30 hours
9	MATH	6
8	PHYSICS	8
8	LANGUAGE	0
42	GENERAL CURRICULUM	42
107	TOTAL	86
124	HOURS TO GRADUATE	124
17	ELECTIVES	38

Since chemistry majors learn calculus in the Mathematics Department, for example, should they not learn computers in the Computer Science Department? The difficulty is that Computer Science departments have different priorities from departments in the physical sciences. Often they are heavily oriented towards business applications. Their vocabulary is different from that of physical scientists; for example a "laboratory" exercise means a software project, whereas to a scientist the term refers to the manipulation of a physical system. At Trinity a student would have to begin with a 3-hour introductory course taught using APL. To get the desired exposure to FORTRAN, he or she would take an additional one hour FORTRAN language lab. With a four hour investment, the students would still have no experience with the computer in a laboratory.

In lieu of a required course, we have integrated

much computer instruction into the Physical Chemistry course, which I teach. Table 2 shows the computer and electronic experience of a typical class of entering Physical Chemistry students. In addition to many of the standard Physical Chemistry experiments on kinetics, equilibrium, etc., the lab includes two sizeable computer modules. In the first semester lab, students having no prior computing experience take a four week module on interactive computing, taught in BASIC on the Datapoint 2200. At the end of this module, students are expected to be able to write a program of a level of complexity comparable to, say, solving the van der Waals equation for V by the method of successive approximations.

TABLE 2
Educational Background of Students Entering Physical Chemistry (Fall, 1982)

TOTAL NUMBER OF STUDENTS	17
Number having had an electronics course	0
Number having had physics beyond General Physics	0
Number having studied any computer language	10
BASIC	5
FORTRAN	2
APL	9
Other	2

In the second semester lab all Physical Chemistry students take a four week module on FORTRAN using the IBM 3031. The major goal of the FORTRAN module is not a proficiency in programming but a familiarity with the total computing environment. There is more at stake than learning the arbitrary syntax rules of an arbitrarily chosen language; for example, when a student understands the reason for carriage controls on output formats, it means he or she understands the status of the printer as a processor independent of the CPU. This is a level of understanding which is not required and therefore is not available when the student is learning APL or BASIC. Interactive languages in general shield the student from many fundamental characteristics of the computer.

Because there is much about FORTRAN that is difficult to understand without a familiarity with its origins in batch/card systems, the FORTRAN module begins with batch processing on cards. Initially the focus is on input/output and FORMATs. This is one of the major differences between FORTRAN and BASIC and a

great source of difficulty to novice programmers. The first project the students attempt is to write a simple program to read and print a card deck. I provide sample card decks containing printer pictures (Snoopy, etc.) to add to the interest. They proceed to a program to reproduce a deck (illustrating the use of unit numbers in I/O statements) and one to print a deck requiring two cards per output line. The focus of the course then quickly changes to the use of a time shared terminal. An important part of this module is a tour of the machine room housing the IBM 3031, so that the students can see the devices that otherwise are just diagrams on a blackboard.
It is manifestly more difficult to teach FORTRAN than BASIC in such a short time. The norm of expected performance is rather modest. In a typical final project, the instructor hands out listings of a FORTRAN printer plot routine. The students type it in and debug their own typing errors, and then write a main program to plot some assigned function.
By the end of the two semesters of lab, the students have had experience with both large and small computers and interactive and noninteractive computing.
Many of these students have had only whatever electronics is offered in General Physics Lab. In Physical Chemistry Lab they take a module on electronics which begins with simple DC circuits. In one experiment they are expected to understand the voltage and current relationships in a potentiometer circuit, which is a simple example of a network. In another they learn the principles of transistor construction and operation. In another they work with transducers and compare digital and analog devices for sensitivity and precision.
Some Chemometrics (the applications of practical mathematics to the treatment of laboratory data) is also presented in Physical Chemistry Lab in the form of a 40 minute lecture once a week. In the first semester I cover simple statistics (Student's t, regression, correlation), curve fitting, propagation of errors, etc. In a typical project focusing on numerical integration as applied to Third Law entropy, one of the methods the students use is a digital graphics tablet interfaced to the Datapoint 2200 to measure the area under a curve. Some experiments require them to use computer graphics facilities to do curve fitting on experimental data, such as an APL spline fit function. In addition they are required to do a term paper using a word processor so that they will be familiar with this important technology.

THE "COMPUTERS IN CHEMISTRY" COURSE

"Computers in Chemistry" (CIC) is an elective advanced laboratory course of two credit hours, requiring a nominal commitment of six hours per week of laboratory time. Figure 1 shows the general areas that might be covered by a course with this title; the dashed line encloses the material chosen for the course at Trinity. The goal of the course is to foster an understanding of computers as opposed to programming (see Introduction). The general outline of the course (see Figure 2) is:

        Digital circuits
        Assembler programming
        Digital interfacing

The course does not cover AC circuits, linear circuits, operational amplifiers, analysis of transistor characteristics, data domain conversion processing or analysis, nor algorithms of any type. A representative semester outline is given as an appendix to this paper.

The students entering this course are primarily chemistry majors who have had Physical Chemistry. Since the Chemistry Department was the first department at Trinity to offer a hands-on minicomputer course, there have been a significant number of engineering students; even now an occasional engineering student takes the course in preparation for the more accelerated and deeper courses in their own department. Because the Computer Science Department has not offered a course with a focus on electronics, I have also had a number of computer science students. The course is taught nearly every semester, either as a regularly scheduled course with the CIC title or as an independent study ("problems") course, sometimes listed officially as a course in the Computer Science Department. The enrollment varies from 1 to 4 each semester.

With such small numbers of students the opportunity to teach the course on an individualized basis is excellent. The laboratory in which the students meet is my research laboratory, located adjoining my office. Most of the teaching in this course is done on a one-to-one basis rather than as formal lectures. This allows the student the opportunity to proceed at his or her own pace and also allows for different backgrounds. The computer science students are already familiar with assembler programming but have very little exposure to electronics. The engineering students have the opposite situation. A placement quiz is given at the

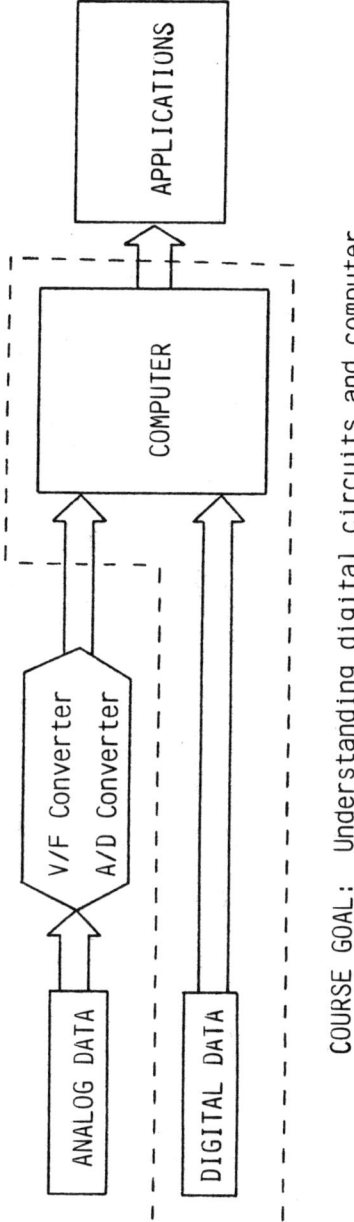

COURSE GOAL: Understanding digital circuits and computer operations at the instruction level.

Figure 1. Areas Covered in "Computers in Chemistry"

REMEDIAL ELECTRONICS REVIEW

Current and Voltage Relationships in a Potentiometer Circuit

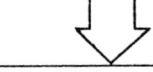

DIGITAL ELECTRONICS

Transistor characteristics
Gates
    NOR gate from two transistors
    Boolean algebra using IC gates
    Voltage relationships in a TTL NAND gate
Digital Counter
    JK flip flops and four LED's
    Convert hex counter to decimal
    Convert display to 7-segment
    Replace the whole thing with an IC
Timing
    One shots

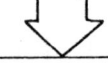

ASSEMBLER

(Run pre-written programs, make small modifications, annotate the programs with comments)
Programs illustrate:
    I/O through the A register
    Interrupts

INTERFACING (Group Project)

Hardware recognition module for 4-bit word

Figure 2. General Outline of "Computers in Chemistry"

beginning of the semester; tnose deficient in electronics background are required to perform some of the experiments done by Physical Chemistry Lab students before proceeding with the course.

This is not primarily an electronics course but a computer course. I have tried to keep the material practical and resist the temptation to add gratuitous theory. Student motivation is seldom a problem: after all, thousands of people make a hobby of it. There are few things more satisfying than building a counter which displays numerals as you push a button, and even more so if you can make a computer read it. There are no quizzes or exams; students are graded entirely on the success of their projects. The required text is "Digital Computers in Scientific Instrumentation" by Perone and Jones (Ref. 1). This book has an excellent treatment of digital electronics and of interfacing.

I am adamant that junior-senior level students should be trained to use professional level resources. Thus if students need technical information on an integrated circuit they are expected to consult the "TTL Data Book" (Ref. 2) rather than a cookbook excerpt in a lab writeup.

The Datapoint 2200 has 16k bytes of memory, a built-in keyboard and CRT, and two cassettes. (See Figure 3.) Through the generosity of Datapoint Corporation we also have the use of a 5-megabyte hard disk and a printer. The instruction set of the 2200 is essentially that of the 8008 microprocessor, although it predates the era of the single-chip CPU. I have had to provide my own teaching materials on the 2200 assembler language and interfacing protocol. I have also written the lab manual for C1C.

The following is a description of the three modules of the course.

The electronics portion of the course covers material necessary to understand a simple digital circuit diagram:
      Logic Levels
      Gates
      Flip Flops
      Counters
      Displays
      One Shots and Timing

It has been very popular to teach such material using patchcord devices which simulate actual electronic components, such as the DEC Computerlab, the Analog Digital Designer or Logicomp. However, while this teaches a good deal of theory very efficiently, it is expensive and still leaves the

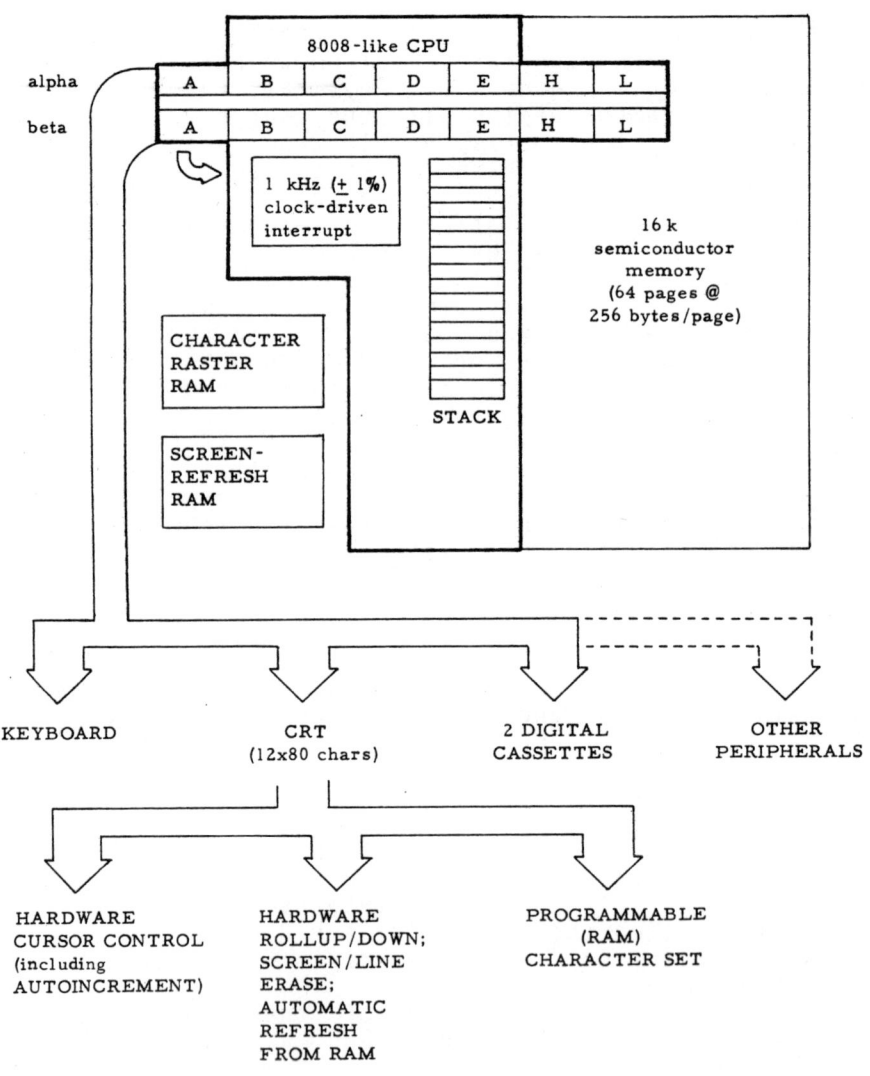

Figure 3. Block Diagram of Datapoint 2200 Computer

students with no ability to repair or construct a device from components. I prefer to use one of the many inexpensive solderless breadboards (e.g., from Global Specialties, New Haven, CT). These devices use ordinary solid wire as patchcords and accept most components directly; thus actual IC's, resistors, etc., may be used. A dual-trace oscilloscope is used in one project to demonstrate timing relationships, an absolutely essential concept. However a digital logic probe covers the majority of troubleshooting situations where an oscilloscope might be used. Such a device uses several LED's to indicate whether a test point in a circuit is constant (high or low) or pulsing; and if pulsing, it roughly indicates the duty cycle (high pulsing low, low pulsing high, or equal); and a memory switch allows one to detect single pulses. Another very useful item is an audible continuity tester. Only TTL IC's are used; they are very tough and reliable, holding up well to student abuse.

After several warmup projects the major task in the electronics module is the construction of a digital counter. The students first make a counter from JK flipflops, strobed by a pushbutton, with a four-LED display. This is an excellent time to introduce them to the hexadecimal number system. Then they design a simple circuit to cause the counter to clear at 10. Next they use a 7-segment LED as the output, and pairs of students combine their counters into two-decade counters. Finally they replace most of the original circuit with a MSI counter / 7 segment LED driver (TTL 74143). This IC requires some programming to select options regarding digit carries, etc.

The goal of the assembler programming module is very limited: the students are not expected to write assembler programs on their own (although some develop an interest in this and go on to do so), but rather to understand several short programs which are presented to them. A number of valuable concepts are conveyed here: the machine architecture, the function of the operating system, the use of the text editor, the difference between operations and pseudooperations, and the use of library subroutines. The students are expected to take a pre-written program, type it in, correct typing errors, and assemble and run it. The program given to them has absolutely no comments or documentation, and the students are graded on the comments they add to the listing, explaining the function of every (or nearly every) line of the program. See the Appendix for the actual projects used.

The last module is on interfacing. The students work in pairs to build and debug a moderately complex digital circuit (about 12 IC's) which communicates with the computer. The students do no designing: the complete circuit diagram is given. The circuit is developed in stages. First a pattern-recognition module is built: a four bit pattern (the S pattern) is set into a dipswitch and then whenever that pattern is written out by the computer a LED is turned on. Then the device is expanded into a translation module: whenever an SPDT switch is set "on" then the output from the computer is monitored and every occurrence of the S pattern is translated to a pattern (the T pattern) set into a second dip switch. If the SPDT switch is "off" then the computer output is passed unchanged.

A project of this complexity is aided by the use of a "super breadboarder" which I designed and built for circuit design and testing. Figures 4 and 5 illustrate this device. A complete description is given in Ref. 3.

A GENERAL PURPOSE INTERFACE DEVICE

The Datapoint minicomputer had no available hardware for laboratory interfacing, and consequently part of the funds originally granted by the NSF were for construction of interfacing gear. With the aid of a designer on the staff of the Trinity Computer Center we built a general purpose interface comprising eight channels of A/D conversion, four channels of D/A, a real time programmable clock and slaved counter, a 16-bit to eight bit multiplexing channel for external digital input, and eight reed relays for external device control.

Complicating the design was the requirement that every control function of the computer be reproduced by manual controls on the machine itself for testing and demonstration purposes, and that every major status and data line be represented on front panel displays. This was done because its primary function was as a teaching device, not merely a laboratory data acquisition system. I felt that it was important for students to learn the internal workings of such a device, for example by having to read the two's complement binary display of the A/D converter output and convert that to a signed floating point value of voltage. Figure 6 is a photograph of this device.

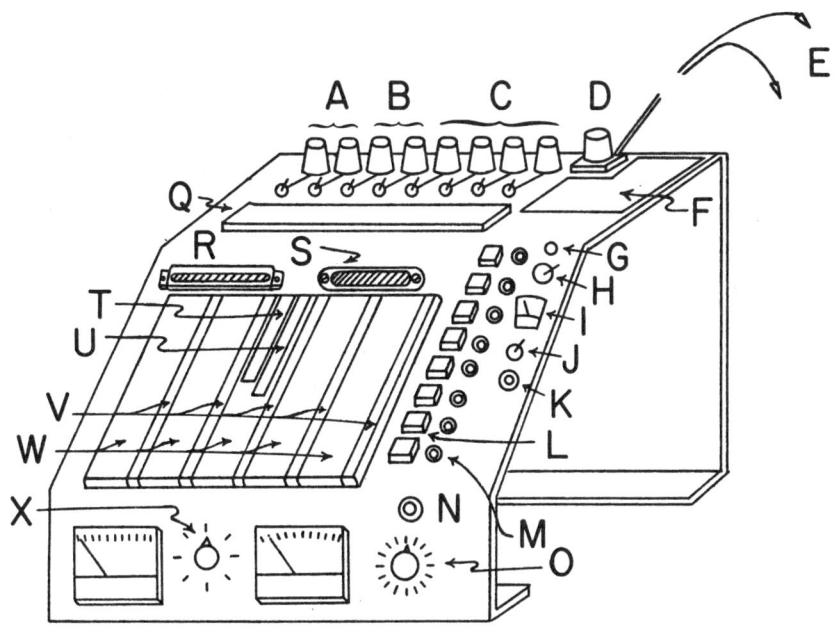

Figure 4. Details of the "Super Breadboarder"
Components: (A,B) splitter input/output; (C) arbitrary inputs or outputs; (D) male BNC plug; (E) socket board probes; (F) power, ground, ammeter inputs; (G) power-on LED; (H) power-on switch; (I) voltmeter; (J) polarity select; (K) voltmeter input; (L) pushbutton connect for status indicators; (M) LED status indicators; (N) TTL clock output; (O) clock frequency select; (Q) socket strip; (R) edge card connector; (S) computer connector; (T) sockets tied to pads of edge card connector; (U) sockets tied to pins of computer connector; (V) bus strips for power, ground; (W) socket strips; (X) ammeter range.

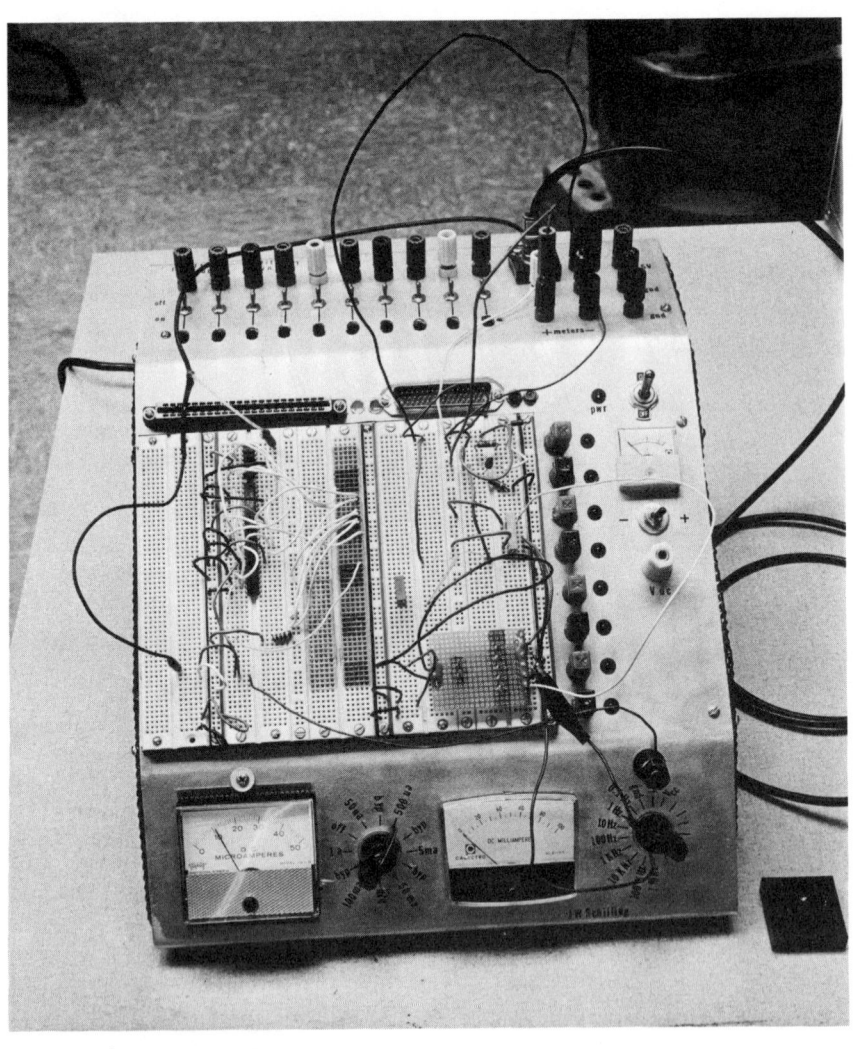

Figure 5. The "Super Breadboarder"

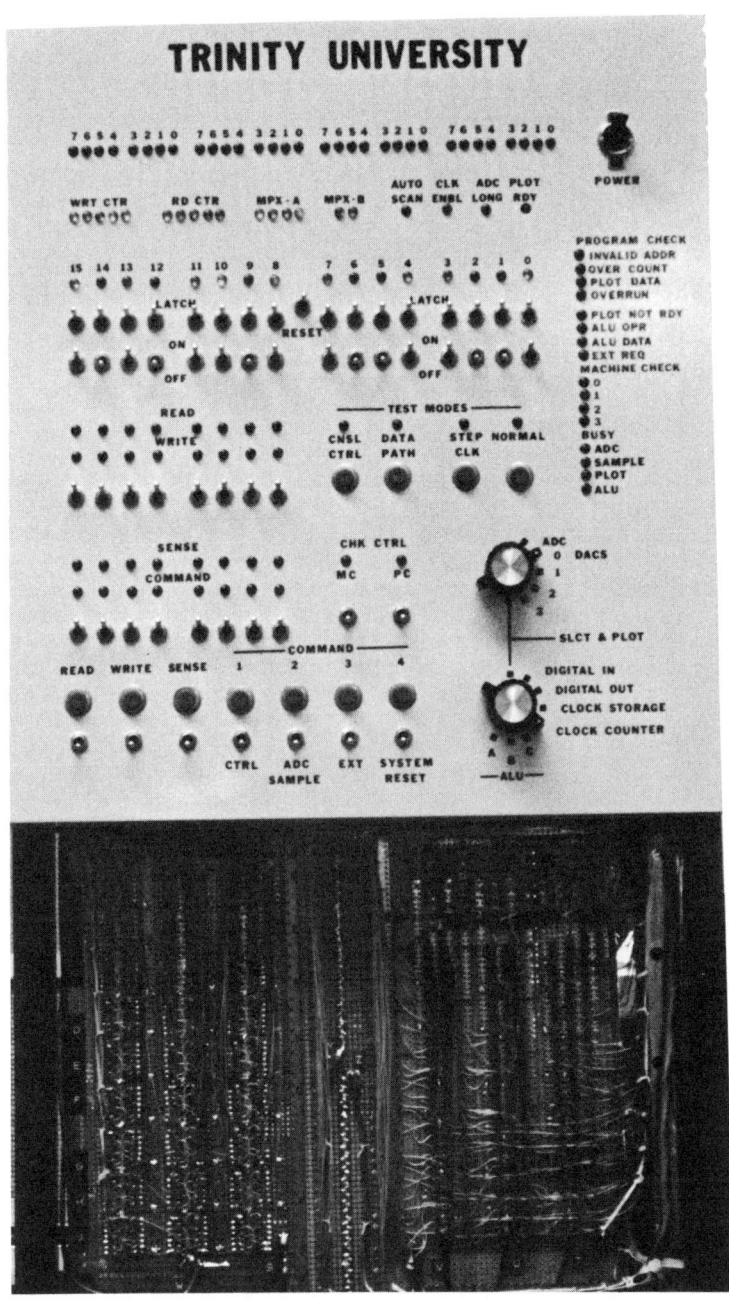

Figure 6. The General Purpose Interface Device ("Schilling's Folly")

CONCLUSIONS AND PROSPECTS

The inclusion of so much computer science in the Physical Chemistry course has distinct disadvantages. Obviously something must be deleted for every inclusion. One of the remedies for this situation is to spread the computer experience throughout the curriculum.

Until recently this has been difficult because of the lack of adequate software and educational documentation. The other members of the staff have not had the interest and expertise to develop such software, and in retrospect they were wise not to try. The microcomputer revolution has brought inexpensive, powerful individual computers for which large amounts of teaching software are available. It has taken the burden of software development away from the individual instructor. Just as significantly, now that a few models of micros are in such wide use, it is becoming easy to find experiments, tutorials, etc, which are specific to the kind of computer you have in your own lab: the burden of producing educational materials is much reduced.

At Trinity we are making a step towards distributing the computer experience away from Physical Chemistry by instituting a one hour "Chemistry Majors' Lab" course, which will be taken by prospective chemistry majors in their freshman year. One of many goals for this course is the early exposure of students to computers and BASIC. This course will be the subject of a multiauthor paper at a later time.

The Computers in Chemistry course will not work at many institutions. It requires a small number of highly motivated students and intimate instructor-student contact. However this is a fairly typical situation at small liberal arts colleges and universities such as Trinity. In this environment it does work. Students occasionally go on to take independent study ("problems") courses in which they work with more advanced applications of computers, both in the laboratory and in straight programming. A great deal of good molecular graphics software (in APL) and several useful devices have come out of such advanced projects.

In teaching CIC I have concluded that a background in "traditional" electronics is not a necessary prerequisite for understanding digital circuits. In fact I would like to suggest a change in the way electronics is taught to science majors. The traditional course begins with DC circuits, moves to analog applications (amplifiers, feedback, etc.), nonlinear devices and AC circuits, and tacks on digital applications at the end. With the pervasiveness of computers, and television hi-fi and telephones going digital, digital electronics should be more than an add-on to a traditional electronics course. A desirable alternative is to turn the syllabus around: after simple DC circuits, move to digital circuits, pulse devices and timing; and then go on to the conceptually more difficult analog and AC circuits. The mathematical tools the students need for digtal electronics (primarily Boolean algebra) are simpler and are very compatible with the principles of programming logic they have already learned. The design of digital circuits within a single logic family is simplified by the modularization of components.

The disadvantages of the Datapoint 2200 are clear from the foregoing discussion: being a business machine it came with no hardware and very little software to support laboratory data acquisition projects. Clearly in today's market one would be foolish to select a business computer for laboratory applications: the availability of appropriate hardware and software are paramount. I would like to make some additional comments on choosing a computer.

It has been suggested that one should use a computer which the students are most likely to find in their eventual workplace. While there are many good criteria for selecting a computer, this is not one of them. If you want students to be able to work with an Apple II when they graduate, then an Apple II is apropriate. But what if they encounter an IBMPC? If one's goal is to familiarize them with the principles by which all computers operate, it does not matter much which computer you use.

In the near future, CIC will be taught using an Apple II. However, during its useful lifetime, the 2200 proved to be very appropriate for teaching a course which involves assembler programming and component- level electronics, and no data acquisition as such. For such a course, simpler is better. The 2200 has a very simple instruction set (8008), which is a subset of that of the 8080, with a few additions to handle built in peripheral functions such as hardware CRT rollup, etc. There are about 20 distinct instructions (not counting different register designations for a given instruction). The addressing scheme is considerably easier to learn than that of the 6502 processor used in the Apple II. It is feasible to teach assembler on such a simple machine to well prepared students in four weeks or so. When the Apple II is used I will teach only a subset of the instructions.

For illustrating the basic features of computers, the Apple II has one sizeable disadvantage compared to the 2200: the Datapoint has a clocked interrupt allowing two levels of time sharing, a great advantage in a laboratory situation where more than one device must be serviced (say, a data generating device and a printer or cassette). The Apple II has no hardware interrupt facility.

In retrospect the construction of the General Purpose Interface (fondly known as "Schilling's folly") was very useful. Its design and construction were valuable learning experiences for myself and for several generations of students of engineering and other disciplines. Today, when other departments have their own computer projects it would probably be impossible to duplicate the effort. Moreover, maintenance has become difficult. Now that inexpensive devices (ADALAB, ISAAC) are readily available commercially it would have less point. Nevertheless, it bears emphasizing that devices designed for teaching, as opposed to routine data collection and transformation, are not being produced. What is needed in education is laboratory data devices that have "windows" built in, through which students can watch what is going on.

The major failing of the computers for chemists curriculum now at Trinity is the skimpy experience the students receive in laboratory computer data acquisition and processing. When the pressure to teach programming languages is removed from Physical Chemistry Lab, some material of this type will be included in that course. In my "dream curriculum" I would have a course dedicated to laboratory interfacing and chemometrics. This course would be offered after Physical Chemistry; assuming that the students have had an exposure to BASIC and FORTRAN, this course would be taught using APL or PASCAL as the language of choice. The students generally would have to learn a new programming language as part of the course material (a standard practice in many computer science courses). The choice of APL may seem strange to those not familiar with it. A few comments about programming languages are in order.

BASIC and FORTRAN are clearly indispensable in today's scientific computing environment because they are so pervasive. However they are "old technology" (some would say "obsolete"). BASIC, in particular, is an abomination: the continual preoccupation with

statement numbers is an unnecessary impediment to programmer efficiency; it lacks local variables, parameterized subroutines and the ability to create subroutine libraries. To be truly well prepared for the "computer age", chemists need a familiarity with some modern programming language.
APL has not lived up to the promise of a decade ago to sweep away FORTRAN, BASIC and COBOL. It is inefficient in its use of machine resources; the succinctness which is one of its strengths makes it inscrutable to all but experienced APL programmers. It puts great burdens on designers of printers and terminals. It does not lend itself to structured programming. Thus it is probably a poor choice for production programs. However each language has its peculiar strengths: the great advantage of APL as a language for teaching algorithms is the power to express complex mathematical operations in a simple way, unencumbered by artificialities resulting from the computing environment. In APL an array (of whatever dimension) is treated as an entity, and only when <u>mathematically</u> necessary does one have to deal with <u>the</u> individual elements. An experienced APL programmer will always be somewhat exasperated with the unnecessary use of FOR or DO loops in BASIC and FORTRAN. (PASCAL is no better than FORTRAN in this regard.) Of all the currently popular languages, APL in competent hands probably requires the least programmer time to achieve a given task. For one-time heuristic programming, it is unexcelled.
There are fads in programming languages: APL in the 1970's and PASCAL in the 1980's. We must be careful to evaluate languages objectively and not be swept away by a bandwagon. Restrictions and control structures encouraging structured programming are touted as PASCAL's strong points, but one can (and many do) write structured programs in FORTRAN. The aids in PASCAL are only marginally helpful for programmers who have developed good habits; for such persons its failings may more than make up for the time saved.
Be that as it may, programming in PASCAL can be a tedious process. PASCAL shares in the major syntactic problems of the whole ALGOL family of languages. For example, suppose an END statement has been misplaced; it is agonizing trying to find which END goes with which BEGIN in a nested set of IF's, WHILE's and ELSE's, especially when complicated by the fact that the presence or absence of a semicolon changes which IF a given ELSE goes with. Indentation only helps if you have made no mistakes; and after several levels of

indentation it loses its helpfulness altogether. (The use of statement numbers a la FORTRAN would have helped eliminate this problem.) Many of the problems with PASCAL occurred because it was designed by Wirth as an example and not intended ever to be implemented on a real machine.

PASCAL's major strong point is in the novel approach to data structures; in this sense it is "modern". But I prefer APL because it requires the students to think differently about their programs than they do in BASIC and FORTRAN. Such a language is worth learning even if the students never use it again. Perhaps this too is a pedagogical bias peculiar to the liberal arts tradition; we care more about how students think about things than how well trained they are to perform certain tasks.

ACKNOWLEDGEMENTS

Financial support was received from NSF Grant GY-11020, the Research Corporation and the Faculty Research and Development Committee of Trinity University. The author gratefully acknowledges the assistance of Professional Engineer Bill McGinnis and the staff of TRINCO (the Trinity University Computer Center), especially Mickey Doherty, Bob Edge, Fred Rodgers, and Dan Laser. Among the students who contributed much to this work are: Rich Davis, Don Kinard, Mike Millner and Matt Yetter. Much of what has been achieved would have been impossible without the cooperation of the Departments of Engineering and Computing and Information Sciences of Trintiy University. The other faculty members of the Chemistry Department at Trinity have been very helpful, enthusiastic and forgiving.

REFERENCES:

1. S. P. Perone and D. O. Jones, Digital Computers in Scientific Instrumentation, McGraw Hill, 1973.

2. TTL Data Book, Texas Instruments Corp., Components Group, Market Communications Dept., P. O. Box 5012, M.S. 84, Dallas, Texas 75222.

3. J. W. Schilling, J. Chem. Ed., 56 A104, 1979.

APPENDIX
Typical Semester Outline of
"Computers in Chemistry"

Week 1
    remedial experiments if necessary;
    construction of a NOR gate from two NPN
      transistors;
    what happens to the truth table when PNP's
      are substituted for NPN's;
    construction of a logic tester from a
      transistor and an LED.

Week 2
    determination of truth table for 7400(NAND),
      7402(NOR), 7486(XOR);
    problems such as forming OR, AND, NAND, and
      NOT from only NOR's;
    formation of OR from NAND's (illustration of
      DeMorgan' theorem);
    use of 7416 and 7417 as LED drivers.

Week 3
    how TTL works.

Weeks 4 and 5 (see text)
    construction of pushbutton-driven counter
      from JK flip flops;
    conversion of hexadecimal counter to
      decimal;
    addition of 7-segment LED display;
    combination of counters to make a 2-decade
      counter;
    use of 74143 IC to make a counter.

Week 6
    use of one shots as pulse stretchers;
    use of oscilloscope and timing
      relationships.

Week 7
    introduction to use of assembler.

Week 8
    an assembler program using external
      subroutines; program adds two floating
      point numbers and displays the result.

Week 9
    assembler program to communicate with
      external device (keyboard/CRT); program

moves CRT cursor around like an
Etch-a-Sketch).

Week 10
    assembler program using clocked interrupts;
    program modulates an audible oscillator
    under keyboard control.

Weeks 11 - 14
    interfacing project (described in text).

GILBERT F. POLLNOW   CHAPTER 10

The MINC-11 as an Exemplary Laboratory Computer for Teaching and Research

INTRODUCTION

    The MINC-11[1] is fundamentally a DEC PDP-11/03 laboratory oriented computer with a 64k MOS CPU designed around the LSI-11/2 processor and LSI-11 bus. It is one of a family of four laboratory oriented computers offered by DEC and is just above the MINIMINC in the heirarchy. Twin eight inch, double density, floppy disks provide over 1 M bytes of random access storage along with a special transparent version of the RT-11 operating system. MINC BASIC provides a rich language of over 400 high level commands or routines which make programming especially easy for the laboratory scientist who does not have the time or expertise to design and assemble his own integrated special purpose computer system. The MINC-11 and its siblings provide bench marks against which future developments in this area may be critically judged as will be demonstrated in this report.
    The system in the Chemistry Department at UW-O includes the following options: one four channel preamplifier, one 12 bit multi-channel A/D convertor, one 12 bit four channel D/A convertor with 6 ma drive capability on each channel, two 16 bit programmable clock modules each with two Schmitt triggers for external

---

MINC-11, DEC, PDP, and MINIMINC are all registered trademarks of the Digital Equipment Corporation.

timing of events, and one D/I module. Standard interfacing hardware includes four serial RS-232 ports, two of which are reserved for the terminal input and a 300 baud printer, and an IEEE-488 parallel bus. Hard copy can be obtained via a Tektronix 4632 video copier if the CRT display is of special interest. Since this copier uses silver based paper, and only 24 lines of code can be copied at one time, an Okidata 82A 120 cps bi-directional printer has been added for program listings and copying of data files. Due to the X-on, X-off protocol used by MINC, the optional 9600 baud serial interface board is required for the printer which with the additional buffer and a revised matching baud rate from the MINC also produces an impressive and sustained printing rate.

The system comes with four master diskettes which can only be copied. The copies are used to operate the system. These include a MINC Master Demonstration diskette which contains the operating system, the Text Editor, a HELP program, and numerous demonstration programs; a System Master which is similar to the previous one but without the demonstration programs; a Diagnostic diskette which can be used to isolate problems in the system, if they occur; and an Applications package containing numerous business and mathematical programs. Eight manuals provide rather comprehensive, user oriented, well indexed, and illustrated instruction. Additional help is available via an 800 DEC phone number during the warranty period and even after.

Courses in which the MINC is used include: Physical Chemistry Laboratory, Polymer Science Laboratory, and a graduate level course. The latter course devotes about two thirds of the semester to numerical methods and graphics using three Tektronix 4051s. Since this course typically includes high school teachers and people from industry, with minimal experience in computing and electronics, the MINC-11 provides an ideal user oriented way of integrating and expanding their experience without having to design their own hardware.

TERMINAL INPUT

Since all communications normally begin with a set of instructions or routines which are part of a BASIC program entered via the VT-105 terminal, a brief discussion of some of its attributes follows. Screen size is 80 columns by 24 lines or 132 columns by 14 lines. Double-width, or double-height double-width characters are selectable on a line-by-line basis with split screen capability. Smooth or jump scrolling, normal or

The MINC-11 as an Exemplary Laboratory Computer 191

reverse video, 7x9 dot matrix characters with descenders, blinking cursor, selectable as either underline or reverse video at the cursor location, a nonvolatile RAM storage of terminal operating parameters and a built-in self-test complete the attribute list of interest to the user. Terminal characteristics can be modified at any time by simply depressing the SET-UP key and following instructions in the manual. Among the characteristics which can be readily changed are: scrolling mode, automatic key repeat, background shading, right margin tone, keyclick, columns per line, and the interlacing (which can be doubled to 480 scan lines for smoother appearing characters).

In addition to the usual ASCII keyboard characters, several special keys are included. The DELETE key is used to correct typing errors. The CTRL key is used in conjunction with certain other keys for control purposes, e.g., CTRL/C is used to interrupt processing at any time. A NO SCROLL key permits the user to examine the scrolling area in the arrested mode. A special set of keys are also provided for convenient control of the text editor which makes a total of 83 keys, counting the space bar. The text editor is sufficiently versatile as a word processor that it was used to prepare the first draft of this report, and is regularly used by the author for most written communications.

Students get their first experience on the MINC as part of the physical chemistry laboratory course in which they statistically manipulate and plot a data set on it for comparison with the Tektronix 4051 with which they have some familiarity from quantitative analysis. Error analysis follows the procedure suggested by Daniels (1), and uses a data set and alternate method of analysis described in detail in the NBS Handbook 91 (2). Students are given the opportunity to examine the effects of the various graphic parameters on their outputs and then to modify the program to fit a set of vapor pressure data to the Clausius-Clapeyron equation. Hard copies for their reports are made on the Tektronix 4632 copier. Typical graphic output and that portion of the program producing it is shown in Figure (1). Tabular output is directed to the printer.

Especially noteworthy in this figure are the GRAPH and LABEL statements which individually generate routines of some complexity but which are transparent to the user. The price paid for this convenience, however, is a loss of some flexibility. Thus labels are always centered in the areas shown, but additional data can be superimposed on the graph via the MOVE CURSOR routine with its line and column specifications, followed by appropriate PRINT statements. The standard

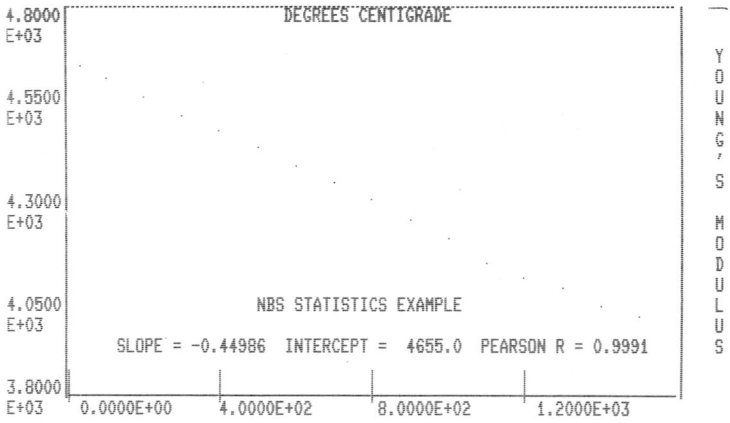

Computer Education of Chemists

```
4.8000 ┌──────────────DEGREES CENTIGRADE──────────────┐
E+03   │                                              │   Y
       │          .                                   │   O
4.5500 │                                              │   U
E+03   │              .                               │   N
       │                                              │   G
       │                    .                         │   '
       │                                              │   S
4.3000 │                                              │
E+03   │                         .                    │   M
       │                              .               │   O
       │                                              │   D
       │                                              │   U
4.0500 │              NBS STATISTICS EXAMPLE          │   L
E+03   │                                              │   U
       │     SLOPE = -0.44986  INTERCEPT = 4655.0  PEARSON R = 0.9991   S
       │                                              │
3.8000 └──────────────┬──────────────┬────────────────┘
E+03    0.0000E+00   4.0000E+02    8.0000E+02    1.2000E+03
READY
```

```
470 S5=SQR(S5/(N-2))
480 REM   COMPUTE PEARSON'S CORRELATION COEFFICIENT, R
490 R=ABS((N*S4-S1*S2)/(SQR(N*S3-S1^2)*SQR(N*S6-S2^2)))
500 PRINT \ PRINT
510 H$="                      NBS STATISTICS EXAMPLE"
520 DISPLAY_CLEAR
530 GRAPH("-HLINES",16,X(0),Y(0))
540 LABEL(,"DEGREES CENTIGRADE","YOUNG'S MODULUS")
550 L=15
560 MOVE_CURSOR(L,10)
570 PRINT H$
580 MOVE_CURSOR(L+2,14)
590 A$="'LLLLLLL##.#####"
600 B$="'RRRRRRRRRRRRR#####.#"
610 C$="'RRRRRRRRRRRRR#.####"
620 D$="'LLLLLLLLLLLLLLLL###.#"
630 PRINT USING A$,'SLOPE = ',A1;
640 PRINT USING B$,'INTERCEPT = ',B1;
650 PRINT USING C$,'PEARSON R = ',R
660 MOVE_CURSOR
670 END

READY
```

Figure 1

graphic area on the CRT can be used to simultaneously display two independent graphs by use of the REGION statement. In either case, the most general form of the graphic routine statement is shown here to demonstrate its remarkable versatility and potential:

GRAPH(option,number,start X,START Y,increment,shade line,graph number,start index)

The MINC-11 as an Exemplary Laboratory Computer 193

In the above statement, only the START Y argument is required with all the rest having default values. Commas are required to the left of the START Y argument but not to the right if the default arguments are desired. A brief description of these arguments below provides a more detailed view of their power:

Options: if the string list is omitted, points and a grid are the default results. Three of the options control the type of symbol used to make the graph: POINTS, LINES, and BRANDS. LINES creates the graph by displaying dots and then connecting them with line segments. BRANDS places short vertical line segments in the positions otherwise indicated by POINTS. GRID automatically includes horizontal lines and tick mark units. If horizontal lines are not desired, the -HLINES is specified as in Figure (1). EXACT produces axis units which are exactly equal to the maximum and minimum input X and Y array values. MOVE enters strip-chart mode in which incoming X values displace the old ones to the left and eventually off the display area.

Number: indicates the number of values to be plotted from the X and Y arrays.

Start X and start Y: specify the initial array elements from which the graph is to be plotted.

Increment: specifies the array element step size to be used. The default value of 1 causes all elements to be plotted whereas a 2 would select every other point.

Shade line: specifies the Y coordinate about which shading is to occur and defaults to the lowest value, i.e., from bottom of the graph.

Graph number: indicates which of two possible figures is to be effected and defaults to graph 1.

A simple example of the split screen possibilities is shown in Figure 2, where the Van der Waals and Ideal Gas isotherms are plotted in regions 1 and 2, respectively. In the event immediate assistance is needed and the manuals are at a distance, typing "HELP" brings forth the menu in Figure 3. If HELP GRAPHICS is requested, the menu in Figure 4 is displayed, which if HELP GRAPH is selected yields the results shown in Figure 5 as described above and illustrated in part via Figure 2.

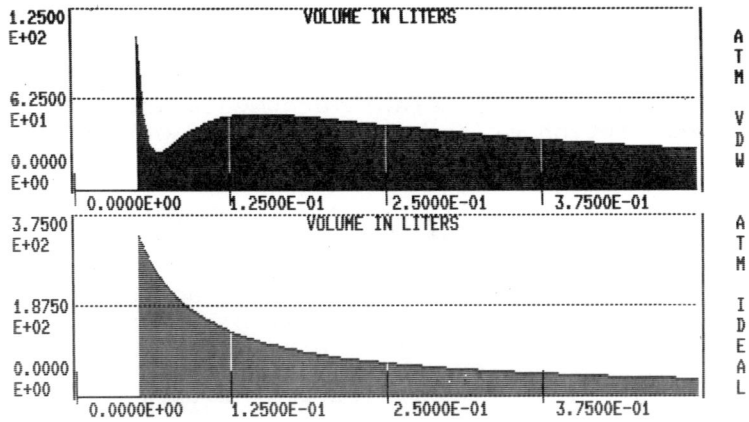

```
READY


READY
LISTNH
100 REM VANIS2: VAN DER WAALS AND IDEAL GAS ISOTHERMS WITH SPLIT SCREEN
110 DIM V(200),P(200),P1(200)
120 T=200 \ R=.08206 \ P0=73 \ V0=.095 \ T0=304
130 A=3*P0*V0^2 \ B=V0/3
140 REM NOTE THAT V > B
150 I=0
160 FOR V=.05 TO .5 STEP 5.00000E-03 \ V(I)=V \ P(I)=R*T/(V-B)-A/V^2
170 P1(I)=R*T/V \ I=I+1 \ NEXT V \ DISPLAY_CLEAR
180 REGION("UPPER",1)
190 GRAPH("SHADE,LINES",I,V(0),P(0),,,1)
200 LABEL(,"VOLUME IN LITERS","ATM VDW",1)
210 REGION("LOWER",2)
220 GRAPH("SHADE,LINES",I,V(0),P1(0),,,2)
230 LABEL(,"VOLUME IN LITERS","ATM IDEAL",2)
240 END

READY
```

Figure 2

```
READY
HELP
HELP
Help is available for the following general topics:

HELP BASIC           Lists all the command, statement, and function names.

HELP GRAPHICS        Lists all the graphic routine names.

HELP IEEE            Lists all the IEEE bus routine names.

HELP LAB             Lists all the lab module routine names.

HELP SETUP           Describes the terminal characteristics.

In addition, help is available for each statement, function, and routine.
To obtain help, type HELP followed by the name of the topic.

READY                                                            -
```

Figure 3

Copyright © 1980, Digital Equipment Corp. All Rights Reserved. Reproduced with Permission.

```
READY
HELP GRAPHICS
GRAPHICS
The graphics routines are:

BARGRAPH         FIND_POINT       MAP_TO_GRAPH     TEXT_INIT
BOX              GET_CHAR         MAP_TO_TEXT      TEXT_LINE
CHAR_MODE        GRAPH            MOVE_CURSOR      VIEW
DISPLAY_CLEAR    GRAPH_INIT       POINT            VLINE
DISPLAY_MODE     GRID             PUT_SYMBOL       VTEXT
DUAL_MOVE        HLINE            REGION           WIDE_LINE
ERASE_GRAPH      HTEXT            ROLL_AREA        WINDOW
ERASE_TEXT       LABEL            SET_BAR
FIND_CURSOR      LIGHTS           SHADE

Help is available for all of the graphic routines. Enter HELP followed
by the routine name.

READY
```

Figure 4

Copyright © 1980, Digital Equipment Corp. All Rights Reserved. Reproduced with Permission.

```
HELP GRAPH
GRAPH(option,number,st-X,st-Y,increment,shadeline,graph no.,st-index)
Argument        Type            Valid values            Default value

option          string expr.    [-]GRID, [-]EXACT,      GRID, POINTS
                                -UNITS, -HLINES,
                                -TICKS, VLINES, INDEX,
                                SHADE, BRANDS, MOVE,
                                POINTS, LINES
number          numeric expr.   0-32767 (integer)       entire array

st-X,           name of         any element of          st-X:   ordinal number
st-Y,           numeric array   any legally             st-Y, st-index: not
st-index        element         dimensioned array                       allowed

increment       numeric expr.   0-32767 (integer)       1
shadeline       numeric expr.   any legal Y             lowest Y value
                                coordinate              in window
graph no.       numeric expr.   1 or 2                  1

GRAPH("INDEX,SHADE",250,X(0),Y(0),2,.5,2,F(0))

READY
```

Figure 5

Copyright © 1980, Digital Equipment Corp. All Rights Reserved. Reproduced with Permission.

ANALOG INPUT

The A/D module is a high quality 12 bit successive approximation type convertor with a biopolar input range of +/-5.12 V with an input impedance of over 100 megohms on each channel and a worst case data acquisition time of 43 micro-sec. Input bias current is a maximum of 40 nano-amps. It is overload protected by a fusible resistor on each channel. In the absence of the pre-amplifier module, up to 16 single ended channels are available without a multiplexer, or up to 64 with a multiplexer. With the Preamp module in the system, channels 0 through 7 are single ended and 8 through 11 are internally and differentially connected with the four preamp channels. A/D channels 0 through 3 have BNC connectors on the front panel and each has its own built-in test voltage and potentiometer which can serve as the input from -5.12 to +5.12 V.

The analog to digital routines include AIN which can collect a single datum, a specified number in a sweep, or an unspecified number of points as a stream of data. AIN-HIST collects a sweep of values and generates a histogram from the data. AIN-SUM collects multiple sweeps and accumulates them for subsequent use in

The MINC-11 as an Exemplary Laboratory Computer 197

signal averaging. Other analog related routines shown in Figure 6 include FFT which performs a fast Fourier transform (or the inverse) on any specified data array. The SCHMITT routine transfers program control to the specified subroutine when an appropriate signal occurs on either of the Clock module Schmitt triggers.

```
READY
HELP LAB
LAB
The lab module routines are:

AIN         DIN_MASK       PAUSE        START_TIME
AIN_HIST    DOUT           POWER        TERMINATE
AIN_SUM     DOUT_MASK      PST_HIST     TEST_BIT
AOUT        FFT            SCAN_BIT     TEST_GAIN
CIN         GET_CHAR       SCHEDULE     TEST_LINE
CONTINUE    GET_TIME       SCHMITT      WAIT_FOR_DATA
COUT        MAKE_BCD       SET_BIT
DIN         MAKE_NUMBER    SET_GAIN
DIN_EVENT   MAKE_TIME      SET_LINE

Help is available for all the lab module routines.  Enter HELP
followed by the routine name.

READY
```
Figure 6

Copyright © 1980, Digital Equipment Corp. All Rights Reserved. Reproduced with Permission.

Figure 7 shows the detailed format of the AIN routine as obtained by simply typing HELP AIN. The example at the bottom of the figure would collect 100 values from the sequence of channels 8,9,10,11, with one sequence every 0.01 second and storing the values in the array V. The precise triggering implied in this demonstration requires a clock module, although other external timing devices can be used via the Schmitt triggers. If only a single clock module is present, its use in the AIN routine makes it unavailable for other elapsed time measurements which then would have to use the system clock with its +/- 1 sec accuracy. If two clock modules are in the system, clock 0 controls all data transfers and clock 1 is reserved for the explicit timing routines: START-TIME and GET-TIME.
   Following a discussion of the successive approximation type of A/D convertor and the Nyquist frequency, students perform a series of laboratory exercises to demonstrate the collection of data in integer and real

```
READY
HELP AIN
AIN(mode,data-name,data-length,trigger,A/D-channel,no.-of-channels)
Argument          Type of Argument    Valid Values           Default Value

mode              string expression   CONTINUOUS,DISPLAY,    standard mode
                                      EXTERNAL,FAST,LINE,
                                      RANDOM,ST2
data-name         numeric variable    -2048 to 2047;         required argument
                  or array name       full-scale values
data-length       numeric expression  >= 1                   1
trigger           numeric expression  0; > 0 to 655.35;      0
                                      1 to 65,535
A/D-channel       numeric expression  0 to 63                0
                  or integer array
no.-of-channels   numeric expression  1 to 64 or channel     1
                                      array length

       Example   AIN(,V(),100,1/100,8,4)
```

Figure 7

Copyright © 1980, Digital Equipment Copr. All Rights Reserved. Reproduced with Permission.

modes using the AIN and GRAPH routines. Figure 8 shows the results of integer sweep of 128 points at a triggered rate of 1000 samples per second using a 50 cycle output from a function generator. The upper region shows the integer input array as a function of the time. The lower region of the output display shows the same data set after converting to volts as shown in line 220 of the program beneath the graphs.

Since the maximum sampling rate of the AIN routine in standard mode is about 1000 conversions/sec, the Nyquist theorem states that the maximum input signal frequency which can be faithfully represented will be one-half the sampling rate or 500 Hz. If the input signal is greater that the Nyquist frequency, the resulting waveform appears as a lower frequency with also a possibly reduced amplitude, a phenomenon called aliasing (3). Figure 9 shows the faithful reproduction of input frequencies of 200 and 400 Hz when sampled at a rate of 1000 conversions/sec. The roughness of the waveforms is due to the linear interpolation used by the GRAPH routines to produce the shaded figures. Figure 10 shows the aliasing which occurs when the sampling rate is reduced to 100 conversions/sec using the same sine wave input as above. In each case a total of 200 points was collected to represent the signal. Incidentally, this figure also illustrates the use of the shade line parameter in other than the default value as utilized

The MINC-11 as an Exemplary Laboratory Computer   199

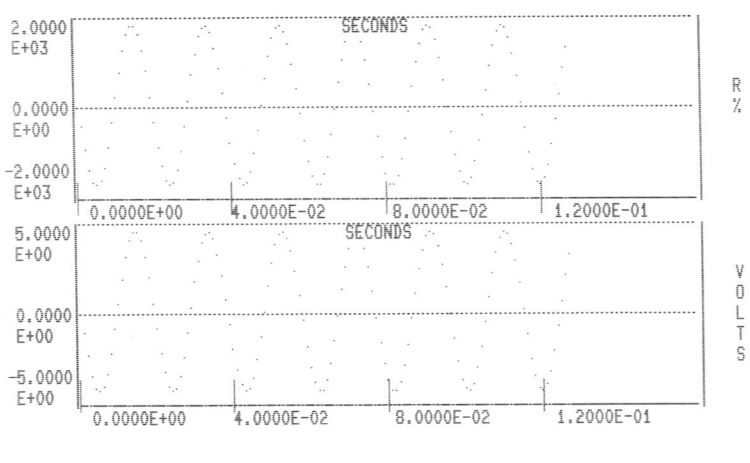

```
READY
LISTNH
100 REM  ADPOL3  ANALOG INPUT IN INTEGER AND REAL MODES
110 DISPLAY_CLEAR
120 PRINT "    THE FIRST PROGRAM WILL READ THE VOLTAGE OUTPUT OF A FUNCTION"
130 PRINT "GENERATOR PROVIDING A SIGNAL OF 100 CPS OR LESS WITH A PEAK TO"
140 PRINT "PEAK VOLTAGE OF APPROXIMATELY 10.24 VOLTS. IT READS 128 POINTS AT"
150 PRINT "AT A TRIGGERED RATE OF 1E-3 WHICH IS THE SHORTEST RATE POSSIBLE"
160 PRINT "USING MINC BASIC. TO EXECUTE THE PROGRAM ENTER ANY NUMBER"
170 PRINT "FOLLOWED BY RETURN WHEN ? APPEARS."
180 INPUT Z
190 DIM C(128),R%(128),T(128),V(128)
200 T0=1.00000E-03
210 AIN(,R%(),128,T0,0,1)
220 FOR I=0 TO 127 \ T(I)=T0*I \ V(I)=5.12*R%(I)/2047 \ NEXT I
230 DISPLAY_CLEAR
240 REGION("UPPER",1)
250 GRAPH(,,T(0),V(0),,,1) \ LABEL(,"SECONDS","VOLTS")
260 AIN(,C(),128,T0,0,1) \ REGION("LOWER",2)
270 GRAPH(,,T(0),C(0),,,2) \ LABEL(,"SECONDS","VOLTS",2)
280 END

READY
```

Figure 8

in Figure 2. Students usually want to see this effect with other waveforms and at other than integral multiples of the sampling frequency. The program below the figures provides for this interaction.

In order to provide students with a hands-on type of creative interfacing experience using the foundations thus far established, it was decided to follow the voltage drop across a load resistance as a function of the

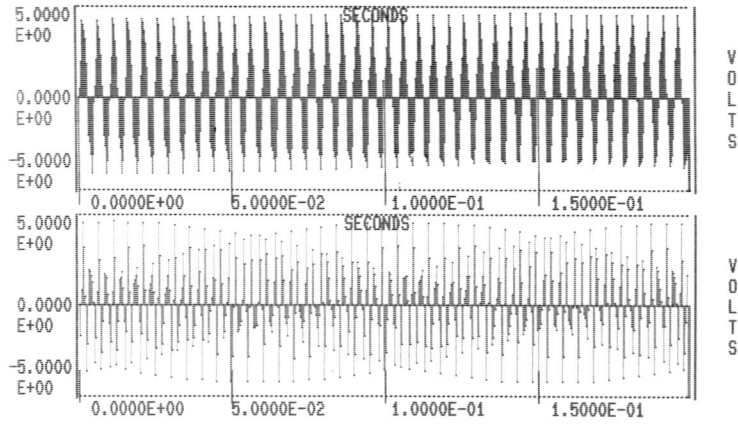

UPPER REGION 200 HZ, LOWER REGION 400 HZ, TRIGGER RATE 1E-3 SEC

Figure 9

current through it, using a 1.35 V mercury battery. Kirchoff's loop rule yields a linear equation for the voltage drop across it when plotted vs. the current. The slope gives the internal resistance of the battery and the intercept, the open-circuit emf. The IR drop across the battery is also equal to the IR drop across the load resistor and the ammeter:

$$E - I*RO = I*(Rl + R2) = V = \text{Voltage drop} \quad (1)$$

where RO is the internal resistance of the battery, Rl is the load resistance, and R2 is the resistance of the current measuring device. R2 is usually negligible with an analog meter, but not necessarily for digital devices which are protected with overload circuits.

Using program handouts from the previous exercises, it is agreed that one of the quickest ways to handle this problem would be to simply add the appropriate AIN statements to the least squares program of Figure 1. Furthermore, the advantage of simulating the behavior with controlled inputs is explained and demonstrated using a Heath Voltage Reference Source to provide linearly decreasing values of V as hypothetical values of the current are input in response to a prompt from the MINC on the CRT. Having thus acquired a feeling for the way in which their program operates, the battery is connected in series with a 750 ohm variable load

The MINC-11 as an Exemplary Laboratory Computer   201

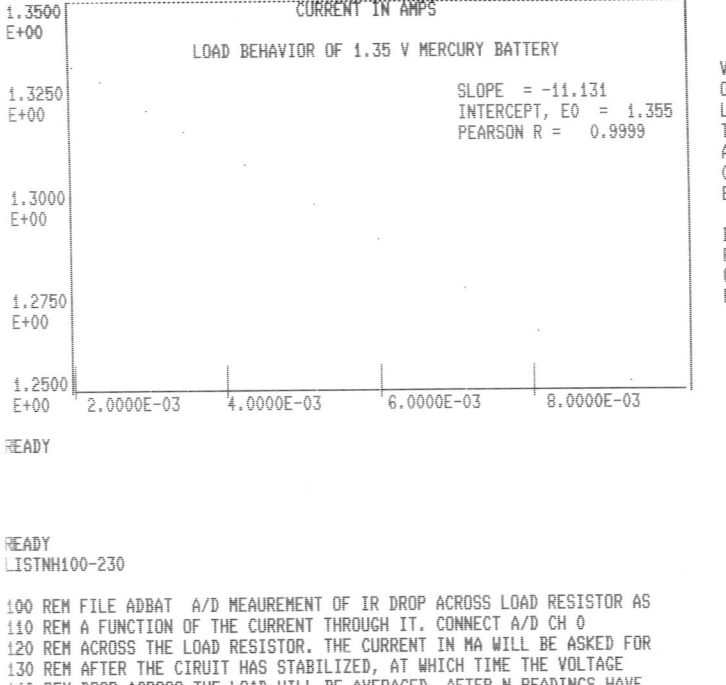

```
READY
LISTNH100-230

100 REM FILE ADBAT  A/D MEAUREMENT OF IR DROP ACROSS LOAD RESISTOR AS
110 REM A FUNCTION OF THE CURRENT THROUGH IT. CONNECT A/D CH 0
120 REM ACROSS THE LOAD RESISTOR. THE CURRENT IN MA WILL BE ASKED FOR
130 REM AFTER THE CIRUIT HAS STABILIZED, AT WHICH TIME THE VOLTAGE
140 REM DROP ACROSS THE LOAD WILL BE AVERAGED. AFTER N READINGS HAVE
150 REM BEEN OBTAINED, A LEAST SQUARES FIT OF THE VOLTAGE DROP VS
160 REM THE CURRENT WILL BE MADE AND PLOTTED WITH STATISTICS.
170 DIM V(5),X(20),Y(20),Y0(20)
180 DISPLAY_CLEAR
190 PRINT "   INDEX       AMPS           VOLTS" \ PRINT \ PRINT
200 N=10 \ FOR J=0 TO N-1 \ INPUT X(J) \ X(J)=X(J)/1000 \ S=0
210 REM TAKE THE AVERAGE OF 5 VOLTAGE READINGS AT EACH OF N POINTS
220 AIN(,V(),5,.01,0,1) \ FOR I=0 TO 4 \ S=S+V(I) \ NEXT I \ S=S/5
230 Y(J)=S \ PRINT J,X(J),Y(J) \ NEXT J

READY
```

Figure 10

resistor rated for 0.8 amps which was selected to keep the current drain to about 10 ma maximum with the prospect for measuring the current via the Preamp along with the voltage drop later. A simple signal averaging routine is included here rather than using the corresponding, but more involved MINC routine AIN-SUM.

For this part of the experiment, a Fluke DMM model 8600A was used to measure the series current in ma when it had stabilized for each resistance setting. The program is intentionally interactive at this point to encourage the student to mentally visualize the flow of the program and its ordered sequence of events. The results of a run and the interactive portion of the program which produced it are shown in Figure 11.

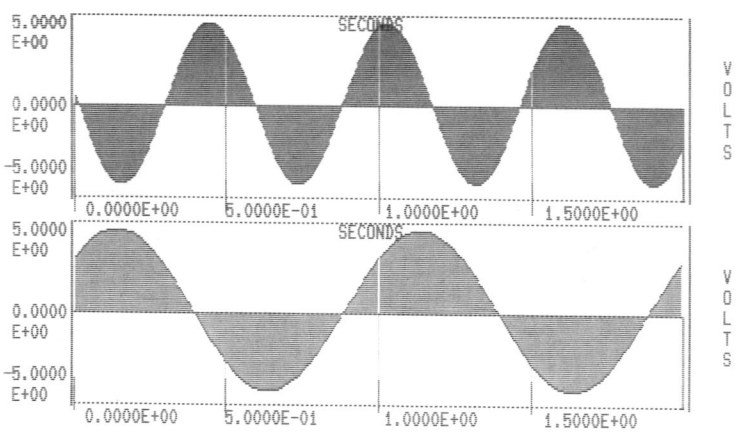

```
READY
        ALIASING: UPPER 200 HZ, LOWER 400 HZ, TRIGGER RATE 1E-2 SECONDS
100 REM  ADNYQ  NYQUIST FREQUENCY DEMONSTRATION
110 DISPLAY_CLEAR
120 PRINT "     THIS PROGRAM WILL READ THE VOLTAGE OUTPUT OF A FUNCTION"
130 PRINT "GENERATOR PROVIDING A SIGNAL OF 100 HZ OR LESS WITH A PEAK TO"
140 PRINT "PEAK VOLTAGE OF UP TO 10.24 V. IT READS 200 POINTS AT ONE"
150 PRINT "POINT EVERY 1E-3 SEC, WHICH IS THE FASTEST RATE POSSIBLE"
160 PRINT "USING MINC BASIC. TO EXECUTE THE PROGRAM ENTER ANY NUMBER"
170 PRINT "FOLLOWED BY RETURN WHEN ? APPEARS."
180 INPUT Z
190 DIM T(200),V(200),V1(200)
200 T0=1.00000E-03
210 AIN(,V(),200,T0,0,1)
220 PRINT " CHANGE FREQUENCY OF FUNCTION GENERATOR" \ INPUT Q
240 AIN(,V1(),200,T0,0,1)
250 FOR I=0 TO 199 \ T(I)=T0*I \ NEXT I
260 DISPLAY_CLEAR
270 REGION("UPPER",1)
280 GRAPH("LINES,SHADE",200,T(0),V(0),1,0,1) \ LABEL(,"SECONDS","VOLTS")
290 AIN(,V1(),200,T0,0,1) \ REGION("LOWER",2)
300 GRAPH("LINES,SHADE",200,T(0),V1(0),1,0,2) \ LABEL(,"SECONDS","VOLTS",2)
310 END
```

Figure 11

Upon examination of the graph and its statistics, it is suggested that the resistance of the battery is untypically high and that perhaps the DMM resistance is significant; so, another run is made with the A/D connected directly across the battery in which case the

The MINC-11 as an Exemplary Laboratory Computer

load now includes the resistance of the DMM. The results are dismally poor since now the voltage drop per step is beyond the resolution of the 12 bit A/D converter.

Replacement of the 750 ohm rheostat with a 56.5 ohm one now allows a sufficiently large IR drop per step to be just within the range of resolution of the A/D. The results are shown in Figure 12 where it is readily seen that the resistance of the battery is now a realistic .636 ohms, and the intercept corresponds well with the label value of 1.35 V. Also, the genuine need for a statistical treatment of the data to the extent shown becomes apparent to the student. At this point, it seems appropriate to let the MINC do all the work and measure the current along with the IR drop. This provides a natural opportunity to introduce the special features of the Preamp module.

ANALOG INPUT VIA PRE-AMP MODULE

The MINC preamplifier provides high quality differential attenuation or amplification of signals internally routed to channels 8 through 11 of the A/D module. Gain can be set via external panel knobs or by means of program arguments. The inner concentric panel knob determines whether voltage, resistance (kilo-ohms), or current (ma) is to be input. The outer knob selects the range for each function, if invariant, or to the program mode. The ranges are: voltage, +/- .01024 to 10.24 V; current, .01024 to 10.24 mA; resistance, 0 to 102.4 K-ohms. In programmed mode, two conversions are made, the first attenuated by 0.5, and the final one at the optimum amplification possible.

In program mode, the SET-Gain (,0,8,3) routine would set channels 8, 9, and 10 to the auto-gain mode indicated by the second argument. The default mode argument, which is the first one in the routine causes the channels specified to sampled in sequential order. Random mode is possible via the RANDOM argument.

Again, the simplest way to incorporate the Preamp into the previous experiment was to leave the IR drop going directly to ch 0 of the A/D, and connect the Preamp in series with the DMM and the load. Initially, the IR drop is measured directly across the battery, with the 750 ohm rheostat as the load, which resulted in IR drops too close together to be resolved. This state of affairs suggests that the load rheostat be reduced while still monitoring the current visually via the DMM in order to protect the Preamp against a possible overload (even though protected). A review of equation (1) suggests that apparently the Preamp has a

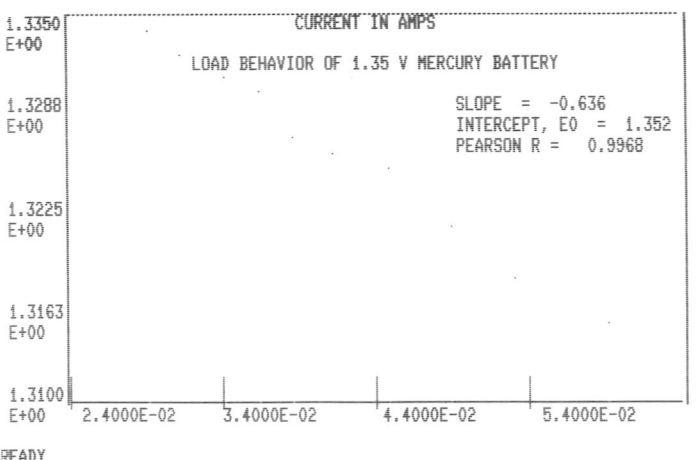

```
100 REM FILE ADBAT2    A/D MEAUREMENT OF IR DROP ACROSS LOAD RESISTOR AS
110 REM A FUNCTION OF THE CURRENT THROUGH IT. CONNECT A/D CH 0
120 REM ACROSS THE LOAD RESISTOR AND PRE-AMP CH C IN SERIES WITH
130 REM BATTERY, LOAD, AND DMM. SET THE PANEL KNOBS ON THE PRE-AMP TO
140 REM PROGRAM MODE AND MA. MAXIMUM CURRENT THROUGH THE PRE-AMP MUST
150 REM BE LESS THAN 10.24 MA. BOTH CURRENT AND VOLTAGE VALUES ARE
160 REM AVERAGED AT EACH SETTING. WHEN THE ? APPEARS AND THE CURRENT
170 REM SHOWN ON THE DMM IS STEADY, ENTER ANY NUMBER AND HIT RETURN.
180 REM AFTER N READINGS HAVE BEEN OBTAINED, A LEAST SQUARES FIT OF THE
190 REM VOLTAGE DROP VS THE CURRENT WILL BE MADE AND PLOTTED WITH
200 REM STATISTICS.
210 DIM X0(5),V(5),X(20),Y(20),Y0(20)
220 DISPLAY_CLEAR
230 PRINT "   INDEX           AMPS              VOLTS" \ PRINT \ PRINT
240 N=5 \ FOR J=0 TO N-1 \ INPUT Z \ S=0 \ S1=0
250 REM TAKE AVERAGE OF 5 CURRENT READINGS AT EACH OF N POINTS
260 SET_GAIN(,0,10,1) \ AIN(,X0(),5,.01,10,1)
270 FOR I=0 TO 4 \ S1=S1+X0(I) \ NEXT I \ X(J)=1.00000E-03*S1/5
280 REM TAKE THE AVERAGE OF 5 VOLTAGE READINGS AT EACH OF N POINTS
290 AIN(,V(),5,.01,0,1) \ FOR I=0 TO 4 \ S=S+V(I) \ NEXT I \ S=S/5
300 Y(J)=S \ PRINT J,X(J),Y(J) \ NEXT J
```

Figure 12

fairly large input impedance even in the current mode. It is suggested that if the IR drop across the battery were made to include the current devices, that meaningful measurements might be made, although the slope now will include the resistance of the DMM and the Preamp. The results are shown in Figure 13. Clearly, the resistance of the Preamp is about 1000 ohms, a somewhat unexpected result from first principles.

The MINC-11 as an Exemplary Laboratory Computer 205

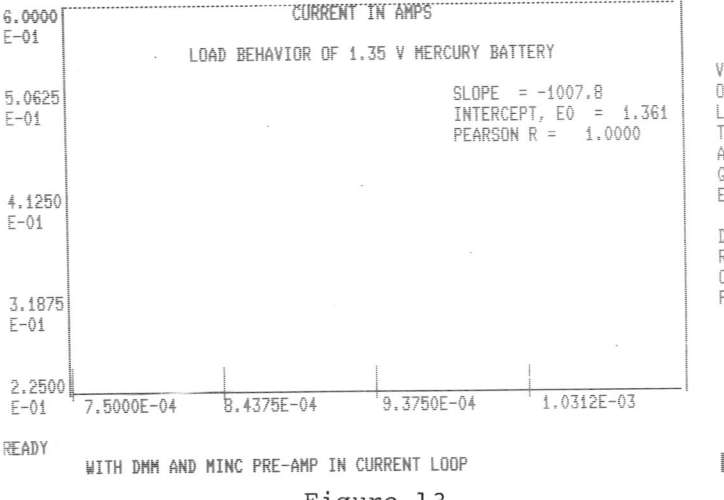

Figure 13

The frustration at not being able to measure the latter case, suggested that by possibly using a 6 volt lantern battery, a large enough IR drop might be produced sufficient to be resolved for the determination ot its internal resistance. This necessitated connecting the IR drop to the Preamp channel 8; so, the SET-GAIN routine was modified as shown above, setting channels 8, 9, and 10 to the programmed auto-gain mode. After modifying the AIN routine to reflect the new channel for the IR input, a number of runs were made. Again, the change in the IR drop between positions of the rheostat were too small to be resolved using the MINC to measure the current. Finally, satisfactory measurements were made by manually entering the currents measured with the DMM, but which would be redundant to illustrate here.

With this experience behind them, students are introduced to differential thermal analysis (DTA) and a vintage Fisher Model 260-P. Although not up to current standards quantitatively, it is electronically simple enough to use as an interfacing project for the MINC. As modified for our purpose, the control thermocouple to the temperature programmer was not interfaced. The platinel differential couple which measures the temperature difference between the sample and the aluminum oxide reference was connected to channel A of the Preamp, and the platinel temperature indicating couple to channel B. Data provided in the original manual for converting the voltage outputs to temperature was fitted to a quadratic equation which was incorporated into the program to convert the analog inputs to degrees Centigrade.

The system clock was used to keep track of the time in order to monitor the actual heating rate versus the supposed rate indicated on the programmer. From 200 to 300 points were collected for each sample and the data stored in virtual array files. This permitted re-examination of any given data set and its graphical representation for further study. Figure 14 shows the results of a typical thermogram, in which the peak temperature of the endotherm happens to agree well with the literature value. In other cases the agreement is rather poor, for reasons related to the intrinsically unsophisticated design. Good pseudo-quantitative results were even obtained in determining the glass transition temperatures of a variety of polymers.

ANALOG TO DIGITAL OUTPUT

The D/A module contains four independent 12 bit converters, each of which can be accessed as a channel by the AOUT routine. Numeric values received via the program are stored in holding registers for each channel converting them to equivalent voltages at the output terminals. The resolution of each converter is 1 part in 4096 or about .025 per cent, which is sufficiently accurate for most laboratory measurements in chemistry. Four concentric panel knobs allow the user to select uni or bi-polar mode via the inner knob, and the range with the outer knob. In uni-polar operation the ranges are: 0. to 5.12 or 0 to 10.24 V. The MINC system can support up to four D/A modules.

Students are introduced to the D/A module after discussing the AOUT routine arguments, noting the similarity to those of the AIN routine described above. Figure 15 gives the on-line version of the AOUT routine produced on the CRT in response to the HELP AOUT menu selection. Apart from the FAST and DISPLAY modes of the AIN routine, the arguments and their default values for the AOUT routine appear the same except for the interpretation of function. For the given example, the AOUT routine sends 512 elements of the array V() in pairs to DAC channels 0 and 1 in an untriggered sweep or burst. When the trigger argument is greater than zero and none of the mode designators specifies an alternative time base, conversions are triggered by the clock module at intervals equal to the value of this argument in seconds. In EXTERNAL mode, conversions are triggered at a periodic rate determined by the external

The MINC-11 as an Exemplary Laboratory Computer   207

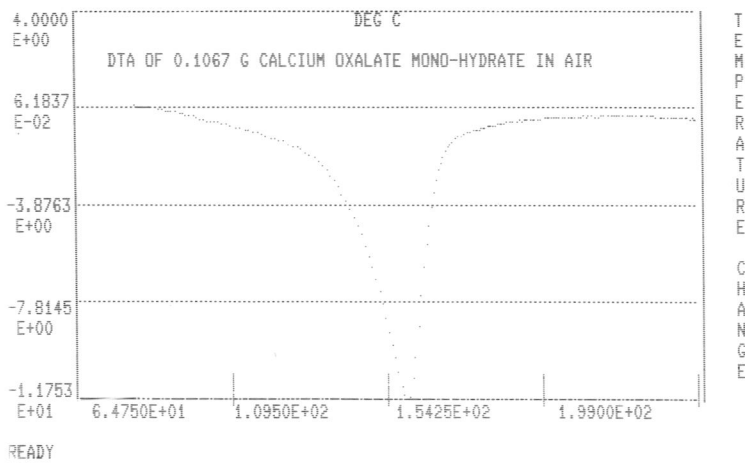

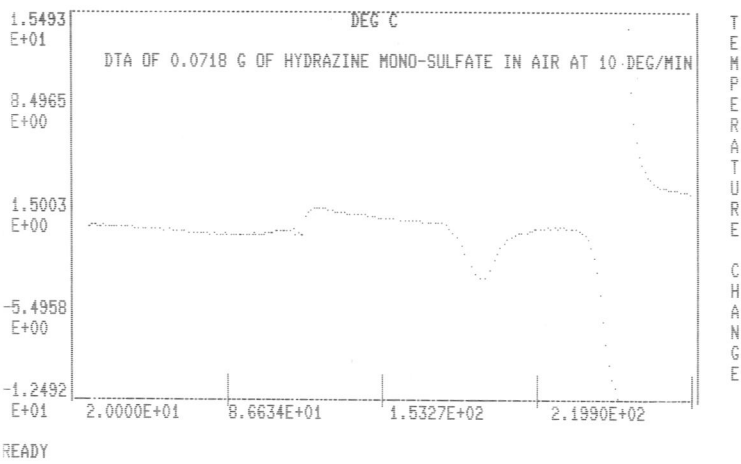

Figure 14

clock connected to ST1 (Schmitt Trigger 1).
  In the laboratory, students review the MINC demonstration program DADEM.BAS in hard copy, and then follow the instructions of the manual in connectng the D/A channel 0 to an analog type D.C. voltmeter set for a full scale deflection of about 6 volts. Execution of the program DADEM then produces a linearly oscillating function whose amplitude increases to a maximum and then halts after the time specified by the user (up to 30

```
READY
HELP AOUT
AOUT(mode,data-name,data-length,trigger,D/A-channel,no.-of-channels)
Argument         Type of Argument    Valid Values          Default Value

mode             string expression   CONTINUOUS,EXTERNAL,  standard mode
                                     LINE,RANDOM,ST2
data-name        numeric expression  -2048 to 2047         required argument
                 or numeric array
data-length      numeric expression  >= 1                  1
trigger          numeric expression  0; > 0 to 655.35;     0
                                     1 to 65,535
D/A-channel      numeric expression  0 to 15               0
                 or integer array
no.-of-channels  numeric expression  1 to,16 or channel    1
                                     array length

        Example  AOUT(,V(),512,,,2)

READY
```

Figure 15

Copyright © 1980, Digital Equipment Corp. All Rights Reserved. Reproduced with Permission.

seconds). An X-Y recorder is then substituted for the voltmeter with the output from the DAC going to the Y-axis. The X-axis is set to a time scale of 2 sec/cm and the Y-axis range switch to 0.5 volts/cm. After plotting the output on the X-Y recorder, the results are analyzed by the students in light of the program statements which actually controlled the output.

Analysis of the program indicates that the output is being sent in the immediate or burst mode without triggering. It is decided to explore the AOUT routine in greater detail with triggered output. It is emphasized again that all the routine really does is to send a set of numbers in the range -2048 to 2047 to the DAC which then converts each number into a voltage governed by the positions of the panel switches. For example, in bi-polar 5.12 volt range: Volts = $I(.0025)$ or the corresponding data value for a voltage in this range is $I = V/.0025$.

The students are next directed and guided to construct a short BASIC program which will generate a sawtooth voltage output to the X-Y recorder, a single point at a time, storing both the elapsed times and the voltages in arrays for graphing on the MINC as well. It is specified that the voltage output is to run from -5.12 to 5.12 V in a linearly increasing fashion starting with a triggered rate of one conversion every 2.0 seconds, repeating for a total of five cycles. Program execution is started at the same time as the X-Y recorder sweep

The MINC-11 as an Exemplary Laboratory Computer 209

(with the range settings at 2 sec/cm and 2 V/cm). At the end of each run, the X-Y time base of the recorder was calibrated with an electric timer.

Examination of the recorder tracing shows that 4.0 sec are required for the MINC to make its first conversion, but from that point on the computer tallied elapsed time agrees with that from the trace, i.e., 118 vs. 119 sec. The triggered rate of conversions is then increased to 10/sec and 100/sec and the experiment is repeated. Analysis of the recordings reveals actual to computer tallied times of 9.4/5.9 and 4.3/.59, respectively. The worst case situation is shown in Figure 16 where the nearly perfect symmetry of the output waveforms provides no evidence of the questionable accuracy of the results. Students are impressed by this apparent problem with the computer in providing an accurately timed sequence of conversions.

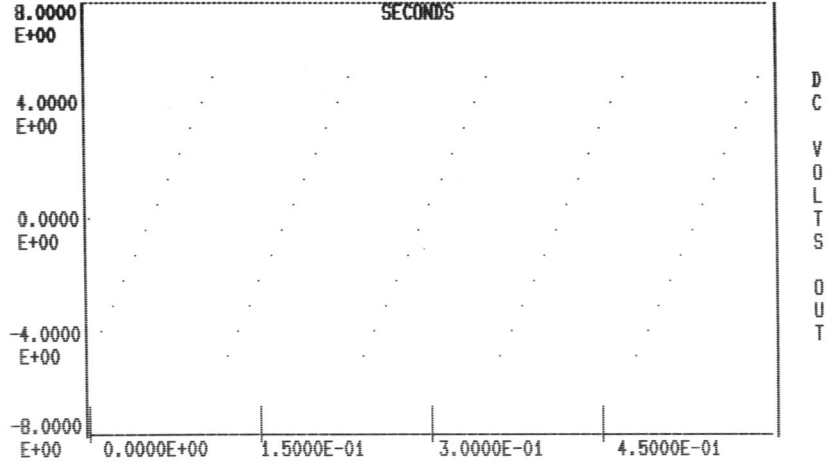

FIGURE 16 D/A DEMO, MINC .59 SEC VS X-Y RECORDER 4.3 SEC, TO = .01

Figure 16

A decision is mutually made to remove the AOUT routine from the bookkeeping loop and instead to make all the conversions with one execution of the AOUT routine with a similar analysis referenced to the X-Y recordings. The actual to computer tallied time ratios are now: 118/118, 5.9/6.4, and .59/.59, respectively. The continuing discrepancy for the intermediate sampling might be due to human error or non-linear behavior of the X-axis time base in the recorder on the 0.5 sec/cm range setting.

The relative ease with which instructive programs of this type can be student generated is shown in Figure 17, where most of the complexity is the result of

```
READY
LISTNH
100 REM  DAPOL2:  D/A DEMONSTRATION WITH SINGLE SWEEP OF DATA
110 REM  CONNECT OUTPUT OF A/D CH 0 TO X-Y RECORDER
120 REM  SET D/A CH 0 TO 5.12 V BIPOLAR OUTPUT
130 DISPLAY_CLEAR
140 DIM I%(256),T(256),V(256)
150 REM  -2048<=FNI%(I)<=2047
160 DEF FNI%(I)=455*I-2048
170 T0=.1 \ N=60
180 PRINT " I           SECONDS         FNI%(I)        VOLTS" \ PRINT
190 FOR J=0 TO 50 STEP 10 \ FOR I=0 TO 9 \ I%(I+J)=FNI%(I)
200 NEXT I \ NEXT J
210 AOUT(,I%(),N,T0,0,1)
220 AOUT(,0,1,,0,1)
230 FOR I=0 TO N-1 \ T(I)=T0*I \ NEXT I
240 FOR I=0 TO N-1 \ V(I)=I%(I)/2048*5.1175 \ NEXT I
250 FOR I=0 TO N-1 \ PRINT I,T(I),V(I) \ NEXT I
260 INPUT Z \ DISPLAY_CLEAR \ GRAPH("-HLINES",N,T(0),V(0))
270 LABEL(,"SECONDS","DC VOLTS OUT")
280 END

READY
```

Figure 17

generating a publishable quality graph of the output. The integer function FNI%(I) provides the linearly increasing sequence of numbers to the DAC in the range from -2048 to 2047. The entire saw-tooth pattern is generated by the single execution of the AOUT routine in line 210. Statement 220 simply returns the output to 0.0 volts. Actual timing for the foregoing analyses in each case included only the interval -2048 <= I%(I) <= 2047. The remaining portion of the program converts the integer array to volts and graphs the data set on the MINC CRT with labels.

Similar considerations, of course, apply to the A/D conversions and great care must be exercised in doing real-time experiments. The MINC, however, allows the user to readily check the accuracy as above, and for time critical measurements at higher conversion rates the system can be upgraded and the programs executed in FORTRAN.

With this background, students are ready for the next project which is to develop a program for following the viscosity of low molecular weight liquids as a function of temperature using the Nametre oscillating

The MINC-11 as an Exemplary Laboratory Computer 211

pendulum viscometer. The program must convert the viscosity/density voltage analog input at an arbitrary temperature to the viscosity/density units required for fitting the data to the linear form of the Eyring equation from which the activation energy and entropy of viscous flow can be determined as described in Daniels (4).

Computationally, this requires determining the calibration function for the thermistor at a number of temperatures between the freezing and boiling points of water. The density-temperature function is taken from the literature. Stirring is to be maintained vigorously except during the 15 second intervals required for the viscometer to reach a steady-state output voltage. Temperature, its reciprocal in deg K, the viscosity-density input, and the logarithm of the viscosity/density are to be tabularly displayed on the CRT as each measurement is made, and the Eyring plot made after all the data have been collected.

To keep the programming as simple as possible at this stage, the sample will be heated to a temperature slightly above that at which measurements are to be made. Upon removing the cup heater, measurements commence when the return key is depressed. By using a fixed period between measurements initially, students readily observe the natural exponential rate of cooling. If the time of each measurement is also stored, a quantitative treatment of the cooling rate is possible. This observation also suggests a simple way of determining specific heats by following the heating rate in an adiabatic calorimeter (or Styrofoam cup). After examining the graph of the data, students can repeat the run and space the points more uniformly by exponentially increasing the time between measurements as the sample cools. Until the program is de-bugged, measurements are preferably made on water to minimize the time of exposure to the sample vapors if adequate ventilation is not possible. Figure 18 shows typical student results on a 70 ml sample of n-butanol using the exponential cooling method.

While the above method works reasonably well as a demonstration, the cooling rate initially is too fast to really assume a steady-state between the sample and the measuring probes; hence, the next student project is to design a programmable proportional temperature controller using the D/A module to control the power to the heater, and the A/D module to monitor the temperature. In the ordinary analog type of proportional controller, the voltage to the heater is reduced as the difference between the sample temperature and the setpoint approaches zero. Since the electrical energy going to the sample holder is the product of the power

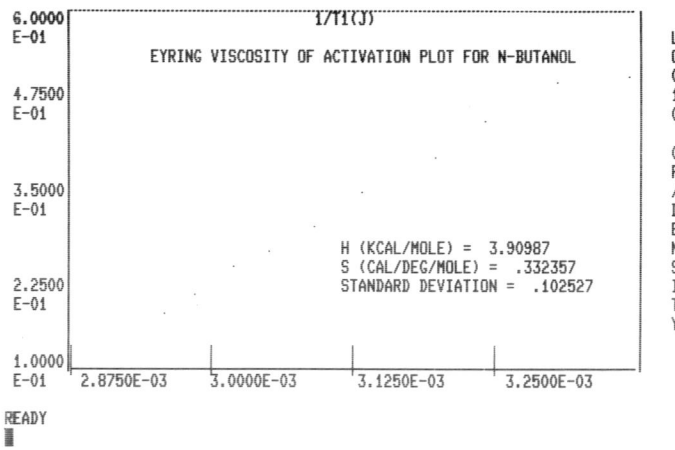

Figure 18

x time, a proportionating effect should be possible by programming the heater "on" time to be proportional to the difference between the set-point and the sample holder temperature, holding the actual power input constant. In case of "over-shoot", the heater "off" time will be kept in proportion to the degree of "over-shoot".

For this project, a cut down tall form 180 ml beaker was wrapped with sufficient 6.5 ohm/ft nichrome wire to yield an approximately 25 watt heater which was then encapsulated to a thickness of about 1.5 mm with a room temperature cure epoxy resin. An iron-constantan thermocouple was then subsequently mounted with epoxy resin on top of the epoxy insulated heater wire, the ends of which were connected to a standard male a.c. plug for use with a Variac. The Variac was set for about 40 volts and was controlled via a 10 amp solid state relay which could be actuated by the 6 ma, 5.12 v output of the D/A converter. Since too fast a switching rate can ruin the relay, an arbitrary "on" or "off" minimum time of 1 sec was required. Temperature of the sample inside the beaker is monitored by means of a calibrated thermistor to +/-0.01 deg C. Stirring was similarly controlled via another channel of the D/A converter as required by the program.

Figure 19 shows the statistical nature of the control process and the sensitivity noise of the thermistor. While print outs of the per cent error from the set-point within the control loop suggest a root-mean-square error of 1.08 percent, observed temperature variations on the plateaus were within +/-0.1 degree Centi-

The MINC-11 as an Exemplary Laboratory Computer 213

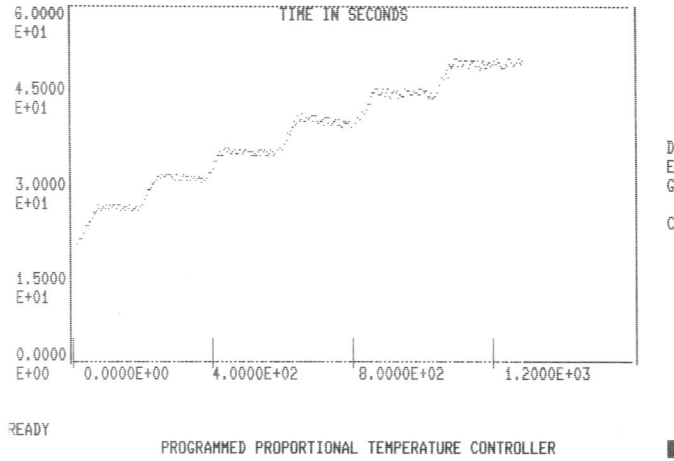

Figure 19

grade as measured with an ASTM yellow-back 0.2 degree thermometer. Similar results were obtained with this programmed controller using a Styrofoam cup fitted with an ordinary cup heater and a Variac setting of 40 volts. The length of time alloted by the program to each step was 3*set-point + 100 seconds. For polymeric liquids, the period per step was increased to 30 minutes due to inadequate or impossible stirring. The effect of signal averaging on the thermistor measurments can also be readily observed and requires only one additional line of code in the program.

USING THE IEEE-488 BUS

The development of low cost microprocessors on a chip has resulted in their utilization in almost all newer instrumentation regardless of cost. However, the ability to communicate with other instruments in digital mode requires a special interface. The RS-232C bus provides for serial flow of data, whereas the IEEE-488 bus is designed for parallel data transfers bi-directionally, e.g., with a computer or other programmable instrumentation.

Communication with other instruments is by means of character strings or messages. Smarter instruments may both talk and listen, whereas less sophisticated ones may only listen. Only one instrument may talk at any time. Any number of listeners is permitted, subject however, to the instructions of the computer controller. Only one controller is permitted in the network. Each instrument on the bus has its own address (a number between 0 and 30) which the computer uses to identify it. Some instruments may have secondary addresses for special purposes.

Part of each instrument's interface is dictated by the IEEE standard and the remainder is determined by the manufacturer. Messages are interpreted by the instrument dependent part of the bus, but they are sent and received through the standard portion of the bus. The computer sends commands to the standard part of the bus which direct its activity but are not sent on to the addressee of the message.

Each instrument on the bus has a vocabulary of characters which it can interpret; so, each message is generally unique to a particular instrument. When a message is complete the talker may send a terminator, e.g., a line-feed or carriage-return. Other forms of termination are also possible depending upon the specific instrument. Further details concerning the IEEE standard can be found in the instrument and computer manuals involved.

Experiments utilizing the IEEE bus have been limited to a GenRad 1658 Digibridge, since it is the only instrument in the Chemistry Department with this capability. This device is a digital impedance meter and limit comparator which incorporates the latest microprocessor technolgy. By virtue of a patented new measurement technique in which a microprocessor computes the desired impedance parameters from a series of 5, 8, or 16 voltage measurements, no user calibration adjustments are ever required. Its basic accuracy is +/-0.1 percent except at the extreme ends of the various ranges.

After a discussion of the essential features of the IEEE bus and the character string components which actuate the various activities of the Digibridge, students are provided with several demonstration programs illustrating the concepts to be examined in the laboratory. The first demonstration invokes the MINC BUSDEM program which unfortunately does not work as written. Discussion of the difficulties reveals the problems of DEC trying to anticipate all possible programmable instrument manufacturers control logic. The last of the demonstration programs, shown in Figure 20 integrates the communication concepts involved and is described in part below.

IEEE-BUS-CLEAR and ALL-INSTR-CLEAR perform the routines suggested by the mnemonics. SET-TERMINATORS(10) designates a line-feed character as the message termi-

```
_ISTNH
100 REM PROGRAM  [RLC4] TO READ GEN RAD 1658 DIGIBRIDGE VIA IEEE-488 BUS
110 IEEE_BUS_CLEAR \ ALL_INSTR_CLEAR \ SET_TERMINATORS(10)
120 C=3 \ N=2 \ SEND("D2S2C1F0L1R4M2X4E1",C)
125 PRINT "      MESSAGE C$               OHMS" \ PRINT
130 FOR I=0 TO N \ TRIGGER_INSTR(C) \ INSTR_TIME_LIMIT(0)
140 RECEIVE(C$,,C) \ R=VAL(SEG$(C$,10,15)) \ E$=SEG$(C$,5,6)
150 IF E$=' O' THEN   \ F1=1 \ GO TO 180
160 IF E$='kO' THEN   \ F1=1000 \ GO TO 180
170 IF E$='MO' THEN   \ F1=1.00000E+06 \ GO TO 180
180 PRINT R,E$,F1*R \ IF I=N THEN 190 \ PAUSE(10) \ NEXT I
190 END

READY
RUNNH

        MESSAGE C$            OHMS

   100          O             100
   .9999        kO            999.9
   .10001       MO            100010
```

Figure 20

nator from the Digibridge which sends both a carriage return and line-feed at the end of each string. Each two characters in the string sent to the Digibridge via 120 set its functions as follows: D2 selects display value; S2, slow measurement mode; C1, series equivalent circuit; F0, 100 Hz measurement frequency; L1, average of 10 measurements; R4, auto-range control; M2, resistance and quality factor to be measured; X4, output only the resistance; E1, disables the manual start switch.

Each time a measurement is to be transmitted, the Digibridge must be triggered as in 130. The argument of INSTR-TIME-LIMIT routine directs the computer to wait indefinitely for the message from the Digibridge. The RECEIVE routine tells the bridge to talk and stores the message in C$. The remainder of the program examines the character string in segments and translates the information into a numerical value which can then be used as a datum for further processing. In implementing these demonstrations a precision resistance box was manually incremented during the 10 second pause at the end of the loop.

Further elaboration of these concepts includes taking a set of ten resistance measurements at regularly spaced time intervals and then plotting the resistance as a function of the time. With this background, students are ready to actually devise an appropriate conductance type experiment. Determination of the equivalent conductance of a weak acid as a function of concentration, as described in Daniels (5), provides a suitable vehicle for their programming efforts and also serves to reinforce the underlying physical chemical principles involved. Typical student results are shown in Figure 21 in which it is readily apparent how the equivalent conductance, $L(I)$, and degree of dissociation, $A(I)$, increase with dilution. The need for extrapolating the equilibrium constant, $K(I)$, to zero concentration is also apparent, and the program is then modified to include to include linear regression extrapolation if the observed errors in the measurement permit.

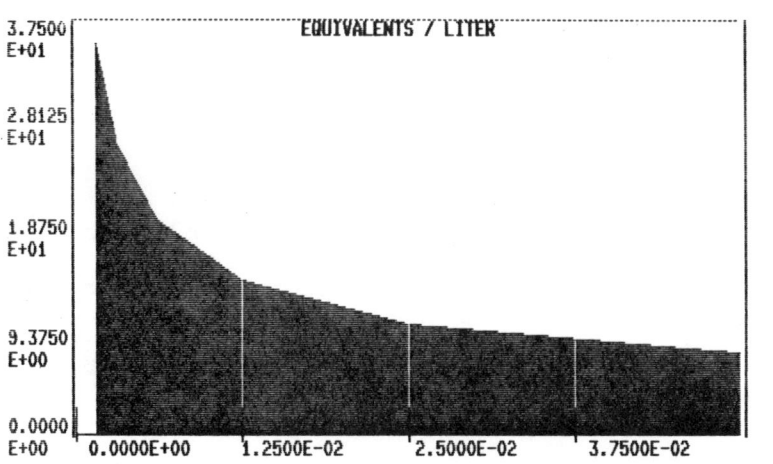

EQUIVALENT CONDUCTANCE OF ACETIC ACID AT 25.00 C

C(I)	OHMS	L(I)	A(I)	K(I)
.05	712.1	7.24578	.0185504	1.75310E-05
.025	1027.6	9.9213	.0254002	1.65496E-05
.0125	1439.4	13.9404	.0356896	1.65111E-05
6.25000E-03	2041.8	19.19	.0491295	1.58651E-05
3.12000E-03	2906	26.0707	.0667453	1.48935E-05
1.56000E-03	4084	35.2804	.0903235	1.39907E-05

EQUIVALENT CONDUCTANCE OF ACETIC ACID AT 25.00 C

Unless the conductivity water used has a very low specific conductance, large errors in the $K(I)$ can be expected at low concentrations of the solute as observed with acetic acid.

While the above experiment provides a suitable programming experience using the IEEE bus for the students, they realize that the batch nature of the individual equilibration measurements wastes a large amount of computer time. This suggests a variety of time dependent phenomena which might be followed on line. One such experiment to determine the rate constant for the hydrolysis of ethyl acetate with hydroxyl ion has been described elsewhere (6). Another somewhat more interesting experiment involving conductance measurements in nonaqueous solvents is the Menschutkin reaction (7). In this nucleophilic substitution reaction between an acyl halide and a tertiary amine in methanol, the conductivity of the solution increases as the reaction proceeds.

By using an excess of the tertiary amine, pseudo first order kinetics can be observed, and by having the MINC make uniformly spaced measurements in time, the Guggenheim method can be incorporated into the program to eliminate the measurement at infinite time (at equilibrium). By following the experimental procedures described in the above reference, the results shown in Figure 22 were obtained. Since the reaction is relatively slow, the system clock can be used to follow the time if a clock module is not available.

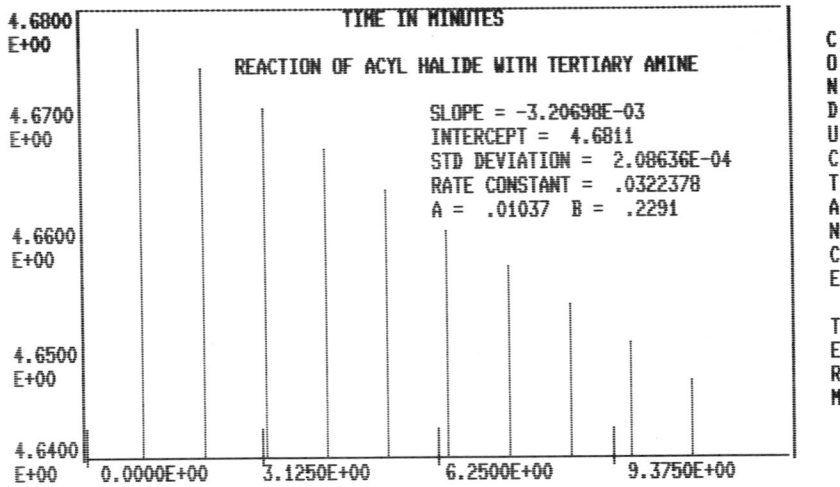

Figure 22

CONCLUSIONS

While this report has attempted to provide some student oriented examples of the use of the MINC-11, it can only hint at the tremendous convenience and power built into the various laboratory routines which make it an ideal laboratory computer system. For those users who might require more memory, double precision arithmetic, or faster execution times a large variety of upgrade packages are available including a real time operating system which permits program development concurrently with data acquisition and FORTRAN IV.

Unfortunately, this kind of hardware and software is relatively expensive. The system at UW-O at the time of this report would cost about $24,000 including the Tektronix 4632 copier. Annual service contracts alone can cost as much as some other complete systems, $2,700/yr just for the MINC. Fortunately, the high quality and reliability of the hardware make it economical to operate without a contract in many cases, (as we do in the Chemistry Department where use of the equipment is intermittent) especially if a DEC service center is near.

The diagnostic diskette which comes with the system also makes it fairly easy to locate the source of most operating problems which might arise. Annual software up-dates cost $320, but some of these prices can be expected to decrease as more competition occurs in the laboratory computer marketplace.

In summary, the MINC-11 represents the best state of the art as a general purpose laboratory computing system for the researcher whose time is expensive, and for the instructor who may not have the expertise to design and build his own interfacing equipment.

REFERENCES

1. F. Daniels, et al., "Experimental Physical Chemistry", Seventh Edition, McGraw-Hill Book Co., Inc., New York (1970) pp. 430-454.

2. M. G. Natrella, "National Bureau of Standards Handbook 91, Experimental Statistics", First Edition, U. S. Government Printing Office, Washington, D.C. (1963) Ch. 5, pp. 1-16.

3. C. L. Wilkins, et al., "Digital Electronics and Laboratory Computer Experiments", First Edition, Plenum Publishing Corp, New York (1975) pp. 141-145.

4. Daniels, et al., op. cit., pp. 157-167.

5. Daniels, et al., op. cit., pp. 175-177.
6. G. F. Pollnow, J. Chem. Ed. 59, 134 (1982).
7. P. W. C. Barnard and B. V. Smith, ibid., 58, 282 (1981).

ACKNOWLEDGEMENTS

This work was supported in part by the UW-Oshkosh Faculty Development Program, and by the National Science Foundation (Grant No. SER 76-16644).

INDEX

ADAPT (Automated Data Analysis, 7
A/D converters, voltage based, 158
A/D module, 196
aliasing, 198
alpha distribution generation, 27
amateurs' literature, 156
analog transducers, 159
APL, 185
artificial intelligence, 9, 84, 119
assembler programming module, 177
aufbau approach to teaching chemometrics, 22

basic cycle time, 123
BASIC-PLUS, 31
Belousov-Zhabotinsky reaction, 114
bidiagonal systems of equations, 134
bit-map graphics, 32
C, 54, 153
central limit theorem Gauss generator, 35
chaining, 127
CHMTRN (chemistry translator), 55
Cehmical Abstracts On-Line, 1
chemical information retrieval, 10
chemical structure, 74
chemometrics, 19, 172
closed loop control, 28
colorimeter, 162
computation intensive problems, 142
computer algebra, 95
  in chemistry course, 173
  interfacing design/practice, 40
  learning, 117
  literacy, 167
  science departments, 169

221

222 Index

CONCISE (infra red spectra evaluation), 55
confidence interval, 28
CONGEN (constrained structure generation), 10
connection matrix, 74
   tables, 7
Cooley-Tucky FFT, 140
countercurrent distribution, 109
Cramer's Rule, 135
curve fitting, 5, 62

data acquisition speed, 155
   intensive problems, 147
   logging, 155
   precision, 155
DENDRAL AI, 10
DIALOG, 40
diatomic rotation-vibration spectroscopy, 39
differential equations, 114
   thermal analysis, 205
digital electronics, 39
   filtering techniques, 36
   impedence meter, 214
   I/O port, 24
dynamic range, 155

eigenanalysis, 6
electronics, 175
ensemble averaging, 19
equivalent conductance, 215

fast Fourier transform, 138
FFTPACK, 138
figure of merit, 86
FORMAC (calculus), 94
FORTH, 153, 164
FORTRAN -77, 53
Fourier transform infrared spectrometer, 6

gas chromatograph, 160
Gauss-Jordan elimination method, 5
   -Laguerre quadrature, 5
gaussian parent population, 36

GELS (evaluation of searching performance), 86
GPTHEORY (group theory), 95
graph isomorphism, 8, 76
   theory, 7, 74
graphical display, 56
Guggenheim method, 217
guided design instruction, 112

hexadecimal number system, 177
Hockney's half maximum performance parameter, 126

IEEE-488 parallel bus, 190, 213
interfacing, 157
instrument interfacing, 154
IRRED (group theory), 95

Kummer functions, 105

Laboratory information management, 59
LaPlace transforms, 116
   and Fourier transforms, 37
LaPlacian operator, 104
LHASA (synthesis analysis), 55
library searching, 73
line shapes, 102
LISP, 54, 102
loop arithmetic errors, 32
   inversion, 142

Marquardt algorithm, 37
matrix algebra, 37
megaflops, 125
Menschutkin reaction, 217
MINC-11, 189
modeling, 62
molecular geometry, 102
   mechanics, 8
Monte-Carlo methods, 6, 117
multi-channel pulse analyzer (software), 35
   -dimensional analysis, 19
   -peak spectra, 36
   -ple linear regression, 37
   -plex spectroscopy, 19

Index

multi-variate statistics, 19
MuMATH (calculus), 95

networking, 70
non-numeric topics, 7
numerical integration methods, 33
   topics, 4
Nyquist frequency, 197

on-line instrumentation, 148
optimization, 8

PAIRS (analyzing infrared spectra), 55
parallel computers, 123
parameter recovery, 37
partial sums problems, 130
PASCAL, 110, 153, 163, 185
pattern recognition, 10, 83
photo-metric titration, 27
   -resistor optical sensor, 162
physical chemistry student, 170
pipelined computers, 123
potentiometric titration, 27
probaility functions, discrete and continuous, 35
problems solving, 107, 154
process control, 159
   devices, 161
PROJECT (group theory), 95
puckering coordintates, 103

random numbers, 117
real time experimentation, 148
REDUCE (calculus), 94
root finding algorithms, 33
roots of polynomials, 112
RS/1 (electronic laboratory notebook), 59
Runge-Kutta Methods, 6

SAINT (calculus), 94
Schmitt triggers, 189
sensitivity noise, 212
single board computers, 156
small sample statistics 27

spectral intensity resolution, 87
library searching, 78
splines enhancement, 33
splitting algorithm, 139
stride, 132
structure elucidation, 80
structured programming, 159
sub-graph isomorphism, 79
   -structure search, 79
super breadboarding, 178
symbol manipulation, 93

Tektronix 4632 video copier, 183
TELAGRAF and DISSPLA graphics software, 57
temperature measurement, 160
textprocessing system (TXTSYS), 52
thermistor, 158, 211
timing relationships, 177
topological indices, 79
   matrix, 74
transcripted data, 148
transducers, 148

useful time, 157

vector calculations, 124
virtual instrument, 164
viscosity, 212

Waterloo BASIC, 159
Wiswesser Line Notation, 7